Fatigue Design

CONFERENCE COMMITTEES

Symposium Chairpersons

J. Solin
G. Marquis
A. Siljander
S. Sipilä

Secretariat

A. Häyrynen
S. Soirinsuo
P. Ukkonen
H. Hänninen
Å. Åvall

International Advisory Board

A. Bakker *The Netherlands*
J. de Fouquet *France*
H. Hänninen *Finland*
H. Kotilainen *Finland*
L. H. Larsson *France*

K. J. Miller *UK*
E. Niemi *Finland*
J. Petit *France*
K. Rahka *Finland*
K. Törrönen *Finland*

Fatigue Design

Edited by
J Solin
G Marquis
A Siljander
and
S Sipilä

ESIS Publication 16

Papers presented at the International Symposium on Fatigue Design organized and sponsored by VTT (Technical Research Centre of Finland) and held in Helsinki, Finland.

Co-sponsored by the European Structural Integrity Society (ESIS) and the Technology Development Centre of Finland (TEKES)

European Structural Integrity Society

Mechanical Engineering Publications Limited
LONDON

First published 1993

ISBN 0 85298 884 2

A CIP catalogue record for this book is available from the British Library.

Typeset by Santype International Limited, Salisbury
Printed in Great Britain by Page Bros. of Norwich

Contents

Preface

Our understanding of the fatigue process has advanced greatly since the days when Wöhler and other early pioneers began studying the effects of cyclic loading on components. However, with the drive towards new materials and more demanding designs, fatigue remains an all too common problem for engineers faced with the design, operation, and construction of engineering components and structures. Information on service loading, material properties, stress analysis, reliability theory, etc. must be thoughtfully combined with a measure of 'engineering judgement' during the analysis and design process. There is often a gap between the information provided by researchers and the knowledge needed by design engineers faced with fatigue problems. Helping to bridge this gap by providing a forum for discussion and the transfer of technology from researchers to practising engineers was one of the goals of an international symposium, Fatigue Design 1992.

The papers in this volume were presented at the FD '92 symposium which was held in Helsinki, Finland on 19–22 May 1992. The symposium was organized by the Technical Research Centre of Finland (VTT), as part of VTT's 50-year anniversary. Approximately 150 engineers and scientists from twenty five countries including seventeen European nations were in attendance.

Topics such as the quantification of service load data for realistic fatigue life predictions, the identification of stress states and failure modes in fatigue, and the residual life assessment in damaged engineering components were among the most common themes put forward for discussion. Attention was also given to the philosophy of fatigue design together with the need for simple yet reliable fatigue life prediction tools to help ensure the adequate fatigue strength of a component or structure at the design stage.

The audience response during the symposium indicated that more case studies in the open literature could be of great assistance to practising engineers. We hope that these proceedings, which contain several papers illustrating direct industrial applications, will help fill this gap and provide new ideas and guidance in the assessment of the adequate fatigue strength of components and structures.

A large number of people contributed time and effort both to the symposium and to this publication. The editors would like to thank the authors for their efforts and the reviewers for their helpful suggestions to the manuscripts. Professor Keith Miller on behalf of ESIS and Louise Oldham of MEP have given invaluable assistance in planning and editing this ESIS publication. Special thanks are due to Professor Kari Törrönen, director of the VTT Metals Laboratory for his financial support and encouragement.

Jussi Solin
Gary Marquis
Aslak Siljander
Sisko Sipilä

Other titles in the ESIS Series

*W. Schütz**

The Significance of Service Load Data for Fatigue Life Analysis

REFERENCE Schütz, W., **The significance of service load data for fatigue life analysis,** *Fatigue Design* ESIS 16 (Edited by J. Solin, G. Marquis, A. Siljander, and S. Sipilä), 1993, Mechanical Engineering Publications, London, pp. 1–17.

ABSTRACT For any fatigue life prediction, be it by experiment or by calculation, the service loads must be known. These can be obtained from previous experience with similar components, by calculation, (this is not, however, very accurate), or by measurement in actual service. The requirements to be met by such measurements are discussed in detail, specifically: (a) what, and how to measure, i.e., stresses or loads, by strain gauge, counting accelerometer etc.; (b) where to measure, i.e., on normal roads, on proving grounds, or on race tracks with greatly increased loads; (c) what counting procedure to use, for example rainflow, range pair, etc.; (d) how to account for variability (scatter) of service loads; and (e) how to account for the (always too) short measurement period etc.

The 'omission' and 'truncation' dilemmas will also be discussed, as well as some problems inherent in experimental fatigue life prediction techniques, like length of the return period chosen and in analytical fatigue life prediction, like accuracy and reliability.

Introduction

Although cyclic service loads and/or stresses have been measured for more than 130 years **(1)** and the cyclic stresses in several different structures had been measured even before the second World War **(2)–(4)**, their underestimation at the design stage has been the cause of a large percentage of the many fatigue problems and failures of machines, structures, and components in the field over the past 40 years. Overestimation of the allowable stresses is the other main type of cause for such failures. The present paper is concerned with the first factor, i.e., the importance of sufficient knowledge of service loads/stresses and how to determine them by measurement and/or calculation.

The knowledge of cyclic service loads, however derived, is necessary for at least two reasons: fatigue life prediction by experiment or calculation is only possible if the service loads/stresses are known and likewise, the collection of loads data over the years for design purposes, standards, building codes etc requires a knowledge of cyclic service loads.

Types of errors in service load data

The causes of such errors are usually one of those listed below.

The service loads are higher than assumed or even measured. The following are typical examples.
– Truck usage in third world countries where the allowable gross weight is considered as allowable pay load.
  Many modern tactical aircraft:

* Industrieanlagen-Betriebsgesellschaft, Ottobrunn, Germany

("In-service usage of the F-18 for both the Canadian forces and the Royal Australian Airforce has shown that the aircraft are being operated in a significantly different manner than assumed for design ...";

"Mainly because the average flight duration is shorter, the rate of fatigue life usage in Australia is more severe than for typical overseas operations of the F-404 engine of the F-18 ...";

"The Swiss usage of the F-5 E/F aircraft is much more severe than previously foreseen for life prediction by the designer of the aircraft ...";

"G-counter records from part of the French Air Force Mirage 2000 fleet have shown that the actual service loads are more severe than those expected in the specifications and applied in the full-scale test ...";

"The accelerations of the fin of the X-29 were much higher than the design conditions and resulted in restrictions ...").

The service loads are known, but the resulting stresses are not calculated correctly. The following is a typical example.
– Aircraft:
("Following an inflight and ground strain survey of the aft wing attachment bulkhead of the F18, it was found that the bulkhead was subjected to higher stresses than predicted by the manufacturer. This area had failed during initial full-scale testing ...".)

The numbers of cycles occurring in service are underestimated or could not be foreseen at design time. The following are typical examples.
– Long range aircraft used for short range flights (all types).
– Thermal power plants of the fifties designed for continuous service, which are now switched on and off several times per day.

The service loads are cyclic, the structure was designed for static requirements. The following are typical examples.
– Early oil rigs in the North Sea designed to Mexican Gulf conditions.
– Generators of a hydro-electric power plant on the Danube, which delivered their output to Eastern countries' power grids, where the frequency varied between 48 and 52 Hz.

The design service life is exceeded later on. The following are typical examples.
– All civil jet aircraft types of the first and second generation (including the Boeing 747) have now exceeded (in some aircraft) their original design life and are still in operation all over the world.

Due to insufficient or incorrect maintenance and repair different and/or higher stresses occur than assumed in design. The following are typical examples.
– Boeing 747 crash in Japan, DC 10 crash in Chicago.
– Two different spindle presses, one the largest in the world.
– US Railroad track system, where German railroad bogies failed, because the stresses were seven times higher than on the German track system.

A few structures of a population are loaded much more severely than the average. The following is a typical example.
– Display flying aircraft.

There are load conditions not considered in design. The following are typical examples.
– Wind turbines: loads due to emergency braking and due to the passage of the blade in front of the tower.
– Pot holes in American roads.
– Six B-47 crashes because the aircraft was designed as a high-level bomber, but later on used for "toss-bombing";
– Aircraft landing gears. ("The most astonishing result is obviously the high loading intensity of the nose landing gear in general and especially in the loading conditions towing and pushing back from ramp ...").

The loads/stresses become more severe during usage. The following is a typical example.
– Military transport aircraft.
   ("It was found that an update of the assumed 1962 usage spectrum of the C-141 was needed by 1968; later, in 1972 it was found that an additional update was needed. The current (1991) operational usage is significantly more severe than that of the second update, because low level training and heavy weight aerial refuelling have increased").

High frequency loads occur, albeit rarely, which were not considered in design. The following is a typical example.
– Hydraulic brake lines of a special vehicle, which came into resonance when the anti-locking brake system operated, which occurred very rarely.

Very high loads/stresses occur in service only a few times, which generate a crack, which then propagates under the normal service loads. The following is a typical example.
– Reduction gear for a 16 1/3 Hz generator which failed after twenty years of service corresponding to about 10^{10} cycles; the high loads occurred during synchronization with the net which happened only eight times in those twenty years.

Consequences of the underestimation of service loads/stresses

The fatigue life of a structure will be shorter if the loads are higher. Although this does not necessarily result in service failures, this is normally the case. There may also be other, entirely unexpected consequences. A case in point is the F-16: it was designed to the slow crack growth philosophy of the USAF Damage Tolerance (5) requirements. The calculated slow crack growth life was > 16 000 flying hours. Divided by the safety factor of two, this gave a life of 8000 hours to the critical crack length of 2 mm, longer than the required life. Therefore, no inspection for cracks was considered necessary during the complete service life, nor was the extremely short critical crack length of 2 mm considered disturbing. However, the service usage in Europe was much more severe than assumed and a recalculation showed the necessity of NDI, combined with the absolute requirement to reliably find cracks <2 mm. Thus the whole maintenance procedure had to be changed and consequently, became much more expensive.

What is known about service loads stresses?

Many branches of industry have measured stresses or loads of their products in service, but not very many results have been published outside the aircraft industry, where at least the general shape of wing lower surface stress spectra or g spectra for tactical (6)(7) and for transport aircraft (6) are well known. However, information on, for example, landing gear (8)–(11) and empennage loads (12) are much harder to find.

The German automobile makers and the LBF, Darmstadt have been measuring loads/stresses on car and truck components all over the world for more than three decades (13)–(17); however, few of their results have been published; moreover stress spectra of automobile components differ more than in any other type of machinery, (i.e., a car driven on a German motorway versus one driven on a cobblestone road). Also there are many cyclically loaded components in a car with completely different spectra, i.e., a crankshaft versus a wheel. The types of load may also be completely different in different countries, for example concerning, i.e., side loads on good roads versus pot holes, i.e., vertical loads on bad roads. Finally there are hard and mild drivers (16)(17).

From the literature, the general shapes of spectra are available for many other components and structures, like bridges (18)–(20), offshore oil-rigs (21), steel mill cranes, steel mill drive systems (22), agricultural machinery (23), and carriages and bogies of rail vehicles (24)(25). However, only *general* knowledge on the shape of the spectrum can be gained from these publications, while *detailed* knowledge such as numbers of cycles per km is required for any specific design. Thus, the measurement of loads/stresses in actual service on each specific design is necessary.

Measurement of loads or stresses?

Normally strain gauges are employed, delivering strain–time histories as a result of the measurement. If the strain gauges are placed in an unnotched section, the proportionality limit will normally not be exceeded and the strains can easily be converted into stresses. By calibration, these stresses can then be converted into forces (or loads) having acted on the component. This is easy if only one load direction occurs in service. If two or three load directions apply like vertical, lateral, and longitudinal loads on an automobile axle spindle, these must be separated. This old problem was only very recently solved by at least two German automobile manufacturers (26)(27).

The knowledge of the loads acting on a component is important for accumulating, over the years with similar components, generalised data for load spectra, load assumptions, standards, building codes etc. If only for this reason, the placement of strain gauges in an unnotched section is by far to be preferred. Strain gauges placed in the notch will only deliver local strains, because in many applications the proportional limit will be exceeded under the higher loads of the spectrum. These elastic–plastic strains are then no longer easily convertible to loads and stresses.

If no strain gauges are employed, then usually g counters at the centre of gravity are used, at least in aircraft, and the resulting g counts have to be converted into loads having acted on the structure, which then have to be converted into stresses. This is not difficult if there is a one to one relationship between accelerations and loads, as in a rigid structure and near the centre of gravity. However, in a flexible structure and far away from the centre of gravity it is difficult. Furthermore, it is impossible for components not stressed by c.g. accelerations, such as landing gears.

Therefore, usually a small percentage of aircraft of one type are strain gauged in many locations of the structure, and the relation between strains and accelerations is calibrated.

Another method for obtaining loads and stresses can be used. The structure, for example of a car, is modelled by a system of springs, masses, and dampers. This model is then 'run' over an artificial road surface in the computer and the response of the structure in many different locations is calculated as stress–time histories. In a similar way, the stress–time histories in steel mill drive systems are calculated (28)(29). The author is very doubtful about the accuracy of this method. It may be good enough for calculations at a very early stage in the design process, indeed it may be the only way to obtain any loads data before the component or structure is available for loads measurement in service. However, if the design process is evolutionary (as is typical for automobile components), experience from earlier, similar designs is of much greater help. This again shows the importance of measuring the service loads acting on the component and not the strains in the notch area.

Evaluation of the measured stress–time history

If at all possible, the stress–time history measured should correspond to normal service usage. This is easy to do for civil aircraft, but is difficult for military transport and tactical aircraft (see above). In some cases it may be useful or even necessary to separate different load cases, like manoeuver loads from gust loads (30) or quasistatic side loads caused by cornering from high frequency side loads caused by surface irregularities. This is usually done by filtering (31) with variable cut-off frequencies.

The stress–time history must then be evaluated (or counted) statistically see Fig. 1, for at least two reasons.

(a) The measurement period will usually not have been long enough to·be used directly in test and for design and the measured stress–time history, therefore, has to be expanded (extrapolated), see below. This cannot be

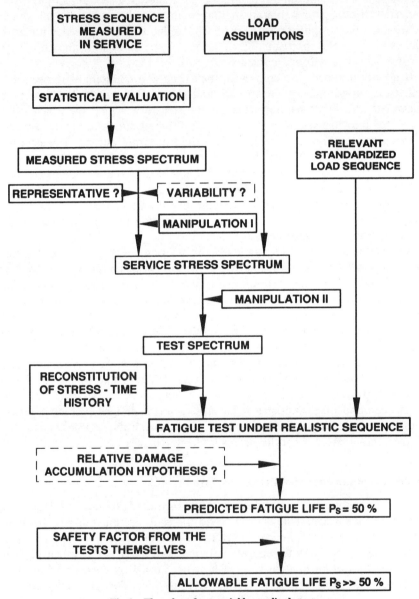

Fig 1   Flow chart for a variable amplitude test

done directly, it is only possible via the result of the counting procedure, the stress spectrum.
(b) The stress spectrum is also needed for comparison with other, previously determined spectra. Only in this way is it possible to accumulate generalized data for load assumptions, standards, and building codes etc, and to recognize if the service loads are, for example, more severe than assumed in the design phase.

There are many counting procedures available, which generally result in as many different spectra from the same stress–time history. The 'Rainflow' procedure, which was developed at about the same time in Japan (32) and in the Netherlands by the NLR (who called it 'range-pair-range') (33) is considered to be the optimum counting procedure, because it counts the stress ranges and the associated mean stresses correctly.

The supporters of the 'local approach' (38) put great stress on the supposition that the hysteresis loops resulting from Rainflow counting are a quantitative measure of fatigue damage. In the author's opinion, this has never been proved. Some authors (34) claim that a Rainflow count cannot be extrapolated, and, therefore, is useless for industrial applications; this is however, not correct. For details on one extrapolation procedure see (35), but other procedures, simpler than in (35) can be employed.

The 'range-pair' counting method counts nearly identical ranges, without regard to the associated mean stress. Some experts consider both counting procedures to be equal. However, Goodmann diagrams for different materials show that the mean stress sensitivity of some is high, while that of others is low (36); hence, generally, both counting procedures are not equal.

One aspect of the Rainflow method needs some further consideration (37) i.e., it will always give the load variation between the lowest trough and the highest peak as the largest range counted cycle. Suppose that this lowest trough occurs very early in the load sequence and the highest peak at the end. If the load sequence in question is a very long one, one may ask whether it physically makes sense to combine these 'remote' occurrences into one cycle. Specifically, in the case of a growing crack, one can imagine that the peak at the end will be seen by a crack which is very different in size and crack tip condition from the one prevailing at the occurrence of the lowest trough.

In other words, it is advisable to restrict the size of the load history on which the Rainflow method is applied at one time. Very long load histories should be split up, each part counted separately and later on added up again. De Jonge suggests in (37) 'one flight' containing several thousand cycles as a reasonable choice of maximum length for the analysis of helicopter rotor loads.

After the counting procedure, the measured stress spectrum is now available, (see Fig. 1). However, before its utilization in the following steps, one should consider what has actually been measured. Is the measured spectrum representative of actual usage? What is the variability of the service stress spectra of nominally identical components? Both of these questions actually refer to the most severe problem several industries have to face in variable amplitude fatigue testing. An automobile maker, for example, does not know with any certainty how his products will be driven by customers; some automobile makers, therefore, try to define the so-called one percent driver, that is the most severe driver out of one hundred for whom the car must be designed fatigue-wise with a certain probability of survival **(39)**.

In the next step the measured stress spectrum must be 'manipulated', see Fig. 1, because in practically all cases the measurement period is too short. Assuming it contains 10^6 cycles (a large measuring and evaluation effort, corresponding to about 2000–4000 km for automobiles or about 10 weeks' continuous measurement time in an oil rig) this is still only about one percent of the required service life of 200 000–400 000 km for an automobile or 20 years for the oil rig, corresponding to about 10^8 cycles in both cases.

The obvious solution is to increase the number of the measured stress amplitudes of all sizes by a factor of 100, see Fig. 2. The next question to be solved is what to do with those events with a probability of occurrence of $P < 10^{-6}$, which are not contained in the measured sequence.

There may be an upper physical limit which cannot be exceeded, for example when an automobile suspension coil spring 'goes to block'. Assuming this happens once during the measurement period of 10^6 cycles, this maximum amplitude occurs one hundred times in 10^8 cycles, but no higher amplitudes occur. If there is no such limit the higher stress amplitudes will have to be added to the service stress spectrum by a meaningful extrapolation procedure,

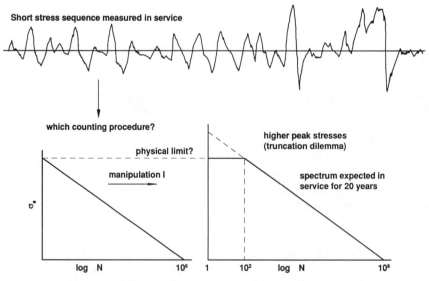

Fig 2   **Extrapolation of the measured stress spectrum (manipulation I)**

like the extreme value method **(40)** which is especially valuable in the case of repetitions of nominally identical events, like loading of a truck. (The truncation dilemma as mentioned in Fig. 2 will be treated below.) We now have arrived at the service spectrum, see Fig. 1.

In place of stress measurements in service, load assumptions have to be used quite often, for example when the complete structure is not yet available for measurements, that is in the development phase.

Such load assumptions are usually based, for example in the automobile industry, on previous experience with similar components, especially if many stress measurements have been carried out before and can be generalized, as mentioned above.

How to use the service stress spectrum for fatigue life tests and calculations

The service stress spectrum arrived at by the procedures described above is, however, not yet a reasonable test spectrum because it has to be manipulated once more in two respects (manipulation II in Fig. 1).

– Large but infrequent stress amplitudes may actually prolong fatigue life due to the beneficial residual stresses they cause. Thus, if the test is carried out with too high infrequent stress amplitudes the fatigue life in test will most probably be unconservative, at least for that percentage of the structures which do not see these stress amplitudes in service. So the correct choice of the maximum stress amplitude to be applied in test, the so called 'truncation dilemma' **(41)** is an important decision. Some experts have suggested that the maximum stress amplitudes in the test spectrum should occur at least 10 times before failure **(42)**.
– Longlife structures, like oil-rigs, ships, trucks, automobiles etc. see about 10^8 cycles during their required life; that is, the service stress spectrum contains 10^8 cycles – too many for an economically feasible fatigue test, because at 10 Hz this means 100 days testing time. So the next question is how best to simulate this large number of cycles. The aircraft industry does not have this problem: the 10^7 cycles a commercial aircraft sees during its service life (and more so the 10^5 cycles a tactical aircraft sees) are usually applied in the full-scale fatigue test.

10^7 test cycles result in a reasonable test time (10 days at 10 Hz). Figs (3a) and 3(b) show four solutions utilized in various industries. Using a spectrum with a more severe shape is one option (see Fig. 3(a) left side) – a typical user is the automobile industry. The problem here is how to read across from the test spectrum to the service spectrum, which requires a relative damage accumulation hypothesis, like relative Miner's rule **(43)**.

10^7 test cycles can also be obtained by omitting small cycles, because this shortens testing time without an increase of the maximum stresses and without

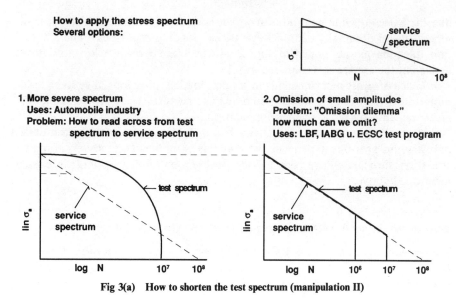

Fig 3(a)   How to shorten the test spectrum (manipulation II)

a change of spectrum shape, see Fig. 3(a), right side. Unrealistically high residual stresses and their possible effect on fatigue life in test are thereby avoided. In a typical straight-line spectrum, this reduction of the number of cycles by one order of magnitude means that all stress amplitudes lower than

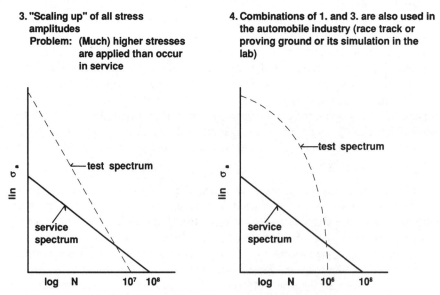

Fig 3(b)   How to shorten the test spectrum (manipulation II)

about 15 percent of the maximum amplitude are omitted; usually they are below 50 percent of the fatigue limit, which has been shown to be a reasonable omission criterion (44)(45). Recent results show that in some cases even two orders of magnitude may be omitted without affecting the result.

If the number of test cycles has to be reduced still further, for example if a low test frequency is thought to be necessary, as in some corrosion fatigue tests, further omission may run into the problem of the 'omission dilemma' (41): The stress amplitudes left out may be near or even above the fatigue limit and the resulting fatigue life in test will be different.

A further option, i.e., the 'scaling up' of all stress amplitudes, see Fig. 3(b), left side, applies higher stress amplitudes than occur in service, with the attendant problems mentioned before.

Using a more severe test spectrum shape plus higher maximum stresses than occur in service is typically the option the automobile makers utilize when they test their cars on race tracks or on their proving grounds in order to shorten testing time, see Fig. 3(b), right side. 5000–10 000 km on such a race track at high speed by test drivers is supposed to be equivalent to more than 500 000 km as driven by the normal customer. Variable amplitude fatigue tests in automobile makers' laboratories also fall into this category. In this case, neither the shape of the test spectrum nor its maximum stress amplitude agree with those of the service stress spectrum and the difficulties of reading across from one spectrum to another are compounded.

Considering the pros and cons of these four options, the author concludes that the omission of small amplitudes is to be preferred.

Most variable amplitude fatigue tests have a fixed sequence which is repeated after a certain number of cycles. The length of this so called return period is critical. On one hand it has to be repeated at least several times, otherwise the various stress amplitudes do not occur in their correct percentages. On the other hand a too short return period means that infrequent but high stress amplitudes are not contained in the test sequence, while they do occur in service and will affect fatigue life. That is a kind of 'truncation dilemma in reverse'. The load spectrum applied in test is thus quite different from that in service.

The above effect is shown in Fig. 4. Again assuming a service stress spectrum of 10^8 cycles, a return period of, say 10^4 cycles has to be repeated 10^4 times and a test spectrum will be applied in which all stress amplitudes above 50 percent of the maximum stress occurring in service have been truncated; such a test will most certainly not give the correct result.

With respect to the return period length the international literature is full of horrible blunders, the worst example probably being the well known SAE-programme (38). The return periods of 1500–4000 cycles which were used because of computer limitations, are just not long enough, as can be seen in Fig. 4. For a fatigue life of 10^8 cycles and a return period of 1000 cycles the maximum stress amplitude occurs one hundred thousand times and a variable

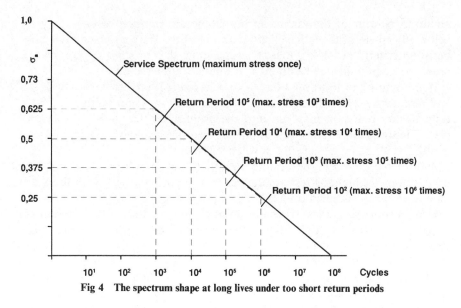

Fig 4   The spectrum shape at long lives under too short return periods

amplitude test to 10^8 cycles is practically a constant amplitude test with this maximum stress amplitude. Moreover, all stresses above 37.5 percent of the maximum stress are truncated; the consequences are discussed above.

In some cases, the length of the return period can be decided quite simply. Tactical aircraft in peacetime are flown very similarly every year for training purposes. So a logical return period is one year and this was, for example, chosen for the 'Falstaff' sequence (46).

The return period is, therefore, a very important decision the test engineer has to take, before a meaningful random fatigue test can be performed.

The test spectrum is now available, see Fig. 1. It has to be reconverted into the test stress–time history for carrying out the variable amplitude fatigue test, the so-called synthesis or reconstitution. If the original counting procedure was 'Rainflow', a Rainflow synthesis has to be used. A number of programmes have been carried out on such techniques over the last few years. There was a coopertive program between the UK, Canada, America and Australia (47) and in Germany two papers were published (48)(49). Broadly speaking, the results of all these programmes show that all these different Rainflow synthesis procedures developed up to now give very similar fatigue and crack propagation test results and also give very similar results to the original stress–time history. So, the reconstitution method seems not to be critical.

Some decisions still have to be taken, for example, when should the maximum stress amplitude occur in the sequence? For this purpose an examination of the test stress–time history is necessary, in order to prevent the accidental occurrence of the maximum stress amplitude right at the beginning of the sequence, which is highly unlikely, for example in a commercial aircraft.

Deterministic or 'abuse' events (like hitting a curbstone), which have not been measured in service, must also be incorporated into the sequence at this stage.

After all the steps described above have been taken, the fatigue test can finally be carried out. If no stress measurements are available, the relevant standardized load sequence, for tactical aircraft for example the 'Falstaff' sequence (46), can be used.

The result of the variable amplitude test is the experimentally predicted fatigue life for a probability of survival of 50 percent, see Fig. 1. However, if the service stress spectrum of the component, whose life is to be predicted is different from the test spectrum, for example if assumptions had been used for the test spectrum and had later been shown to be incorrect by measurements in service, an additional fatigue life prediction by calculation must be carried out, see Fig. 1, in order to read across from the test spectrum to the service spectrum. This is usually done by employing one of the relative damage accumulation hypotheses (43).

The safe fatigue life for the high probabilities of survival required for real components is determined using a statistically derived safety factor, which can be taken directly from variable amplitude tests. Usually only a small number of variable amplitude tests on nominally identical components under identical test conditions are performed. The confidence for the safety factor derived from these few tests is, therefore, low.

It is a much better idea to use previous experience from many variable amplitude tests on similar components and to derive different safety factors, for example, for machined surfaces, cast or forged surfaces or welded joints; some relevant scatter factors are presented in (50), derived from previous experience of thousands of fatigue tests.

The flow chart for a fatigue life prediction by calculation according to the nominal stress concept is shown in Fig. 5. The steps to be taken are the same as in Fig. 1 up to and including the service stress spectrum. However, here a complete S–N curve with the component in question has to be obtained, in the general case even several S–N curves with different mean stresses. By employing a suitable damage accumulation hypothesis, usually Miner's rule, the fatigue life is calculated, again for a probability of survival of 50 percent and the allowable fatigue life calculated for $P_s \gg 50$ percent as in Fig. 1.

The practice of using an S–N curve with a $P_s > 50$ percent to calculate an allowable fatigue life, as used in several standards, for example (51)(52) is fundamentally wrong in the author's opinion. The scatter in fatigue life under constant amplitudes does not correlate at all with that under realistic, variable amplitudes, as shown by hundreds of tests with actual components. It is therefore, necessary to proceed as described above, i.e., to use scatter factors obtained from variable amplitude tests.

The accuracy of fatigue life predictions using Miner's rule and the nominal stress concept generally is poor (53)–(59) for specimens and even worse for actual components (60), where damage sums to failure of 01 and less are

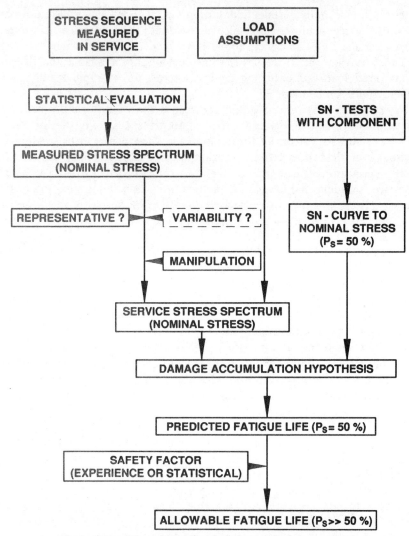

Fig 5 Fatigue life prediction by calculation (nominal stress concept)

quite common. This accuracy could not be improved by variations of the original Miner concept (60), neither by the so-called 'elementary' Miner-Rule, nor by employing a reduced fatigue limit nor an S–N curve with a reduced slope below the fatigue limit, as in (51)(52).

The fatigue life can also be predicted by the so-called 'local approach' (38). It has been shown by a large cooperative programme in Germany (61) to be even less accurate than the nominal stress approach, notwithstanding innumerable (and unfounded) claims in the international literature. This may in part be due to the inaccuracy of Miner's rule itself which, in the end, is also employed in the local approach assuming a damage sum at crack initiation of 1.0.

Conclusions

If the fatigue life of a component is to be predicted with any accuracy, the service stress spectrum *must* be known by measurement *and* a variable amplitude test with this spectrum *must* be carried out. Anything less is at best a makeshift solution which may give highly unconservative results and lead to early fatigue failure, as shown by experience with many different structures and components.

References

(1) WÖHLER, A. (1858) Bericht über die Versuche, welche auf der Königl. Niederschlesisch-Märkischen Eisenbahn mit Apparaten zum Messen der Biegung und Verdrehung von Eisenbahnwagenachsen während der Fahrt aufgestellt wurden, *Z. Bauwesen*, **8**.
(2) KLOTH, W. and STROPPEL, TH. (1936) Kräfte, Beanspruchungen und Sicherheiten in den Landmaschinen, *VDI-Z 80*, **4**, 85–92.
(3) SCHÄCHTERLE, K. (1933) Die Bemessung von dynamisch beanspruchten Konstruktionsteilen, *Bauingenieur*, **14**, 239–242.
(4) KAUL, H. W. (1938) Statistische Erhebungen über Betriebsbeanspruchungen von Flugzeugflügeln, Jahrbuch der deutschen Luftfahrtforschung.
(5) N.N. (1974) Military specification: airplane damage tolerance requirements. MIL-A-83444 (USAF).
(6) TAYLOR, J. (1965) *Manual on aircraft loads*, Pergamon Press Oxford.
(7) (1984) Operational loads data, AGARD Conference Proceedings No. 375.
(8) (1990) Landing gear design loads, AGARD Conference Proceedings No. 484.
(9) BUXBAUM, O. (1982) Landing gear loads of civil transport airplanes. Eighth Plantema Memorial Lecture. *Aircraft fatigue in the eighties. Proceedings of the Eleventh ICAF Symposium* (Edited by J. B. Jonge and H. H. van der Linden), National Aerospace Laboratory NLR, Emmeloord, Netherlands, p. 0/1–36.
(10) BUXBAUM, O., KLÄTSCHKE, H., STEINHILBER, H., and CUNY, J. J. (1991) Loads at the nose landing gears of civil transport aircraft during towbarless operations, *Aeronautical fatigue: key to safety and structural integrity. Proceedings of the Sixteenth ICAF Symposium* (Edited by A. Kobayashi) EMAS, Warley, UK.
(11) BUXBAUM, O., LADDA, V., and ZASCHEL, J. M. (1982) Betriebslasten am Bug- und Hauptfahrwerk eines Flugzeuges vom Typ Airbus A300 B2 während des Einsatzes bei der Deutschen Lufthansa, LBF-Bericht Nr. 3691.
(12) BUXBAUM, O. (1972) A relation between measured CG vertical accelerations and the loads at the T-tail of a military airplane, AGARD-Report No. R-597.
(13) SVENSON, O. and ERKER, A. (1951) Kraftmessung bei wechselnder Beanspruchung. ATM-Blatt V 131-1.
(14) BAUTZ, W., GAßNER, E., and SVENSON, O. (1951) Beanspruchungsmessungen und Betriebsfestigkeits-Versuche an Fahrzeug-Bauteilen ATZ 53 Nr. 11, 286–287.
(15) SVENSON, O. (1955) Untersuchungen über die Größe und Häufigkeit der Betriebskräftezwischen Zugwagen und Anhänger, Deutsche Kraftfahrzeuforschung, Techn. Forschungsbericht Nr. 1.
(16) GÜTHE, H. P. *et al.* (1987) Bewertung der Beanspruchungsstreuung aus gemessenen Kollektiven, *Automobiltechnische Zeitschrift* **89**.
(17) NALEPA, E. *et al.* Mehrparametrische Beanspruchungsanalyse an Automobilbauteilen, Mess- und Versuchstechnik im Automobilbau, VDI Berichte 632.

(18) ZASCHEL, J. M. (1982) Zur Verkehrsbelastung von Stahlbrücken, *Der Stahlbau*, **51** 58–59.
(19) PFEIFER, M. R. (1982) Verkehrslasten und Beanspruchungen von Straßenbrücken, *Ermüdungsverhalten von Stahl- und Betonbauten*. Internationale Vereinigung für Brückenbau und Hochbau, Zürich, pp. 857–864.
(20) SEDLACEK, G. and JACQUEMOND, J. (1984) Herleitung eines Lastmodells für den Betriebsfestigkeitsnachweis von Straßenbrücken, *Forschung Straßenbau und Straßenverkehrstechnik*, Vol. 430, Bundesminister für Verkehr.
(21) SONSINO, C. M., KLÄTSCHKE, H., SCHÜTZ, W., and HÜCK, M. (1988) Standardized load sequence for offshore structures, Wash 1-LBF-Report FB 181, IABG-Report TF-2347.
(22) SVENSON, O. and SCHWEER, W. (1960) Ermittlung der Betriebsbedingungen für Hüttenkrane und Überprüfung der Bemessungsgrundlage, *Stahl und Eisen* **80**, 79–90.
(23) RENIUS, K. Th. (1976) Last- und Fahrgeschwindigkeitskollektive als Dimensionierungsgrundlagen für die Fahrgetriebe von Ackerschleppern, *Fortschritt-Ber. VDI-Z*, **49**.
(24) KRETSCHMER, R.-M. (1982) *Beitrag zur Synthetisierung von Lastkollektiven zur Bemessung der Laufwerke von Schienenfahrzeugen*, PhD Thesis, University of Hannover, Germany.
(25) LÖWE, H.-J. (1987) *Verfahren zur Bestimmung synthetischer Kraftkollektive für Reisezugwagendrehgestelle in Abhängigkeit der Bauart*, PhD, Thesis, University of Hannover, Germany.
(26) KÖTZLE, H. (1984) Mehrkomponentenmessung am PKW-Radim Betrieb, Vorträge der 10. Sitzung des AK Betriebsfestigkeit, Deutscher Verband für Materialforschung und -prüfung.
(27) LOH, R. and NOHL, F. W. (1992) Vielkomponentenradmessnabe *ATZ 94* 1, 44–53.
(28) MÜLLER, V. *et al.* (1984) Anleitung zur rechnerischen Ermittlung von Bauteilbelastungen in Antriebssystemen, Bericht ABF 24, Verein Deutscher Eisenhüttenleute, Düsseldorf.
(29) WÜNSCH, D. (1988) Lastannahmen durch Simulation im Schwermaschinenbau, Vorträge der 14. Sitzung des AK Betriebsfestigkeit, Deutscher Verband für Materialforschung und -prüfung.
(30) BUXBAUM, O. and STEINHILBER, H. (1989) Description and reconstitution of manoeuver landings, *Aeronautical fatigue in the electronic era. Proceedings of the Fifteenth ICAF Symposium*, (Edited by A. Berkovits), EMAS, Warley, UK, pp. 43–63.
(31) KLÄTSCHKE, H. and STEINHILBER, H. (1985) Trennung überlagerter Beanspruchungs-Zeit-Funktionen durch Filterung mit variabler Grenzfrequenz, LBF-Bericht Nr. TB-174.
(32) MATSUISHI, M. and ENDO, T. (1968) *Fatigue of metals subjected to varying stress*, Japan Society of Mechanical Engineers.
(33) DE JONGE, J. B. (1969) Fatigue load monitoring of tactical aircraft, NLR-Report TR 690330.
(34) BUXBAUM, O. (1992) *Betriebsfestigkeit. Sichere und wirtschaftliche Bemessung schwingbruchgefährdeter Bauteile*, Second Edition, Verlag Stahleisen mbH, Düsseldorf.
(35) PETERSEN, J. and KRÜGER, W. (1986) Experimenteller Lebensdauernachweis für Kfz-Komponenten auf der Basis von rekonstruierten stochastischen Beanspruchungen, VDI-Berichte 613.
(36) SCHÜTZ, W. (1974) Schwingfestigkeit von Werkstoffen, VDI-Berichte 214.
(37) DE JONGE, J. B. (1983) The analysis of load-time histories by means of counting methods, Helicopter Fatigue Design Guide; AGARDograph No. 292.
(38) (1977) *Fatigue under complex loading. Analyses and experiments*, (Edited by R. M. Wetzel), Society of Automotive Engineers, USA.
(39) WIMMER, A. and PETERSEN, J. (1979) Road stress resistance and light weight construction of automobile road wheels, SAE Paper 790713.
(40) BUXBAUM, O. (1968) Verfahren zur Ermittlung von Bemessungslasten schwinggefährdeter Beuteile aus Extremwerten von Häufigkeitsverteilungen, *Konstruktion*, **20**, 483–489.
(41) CHRICHLOW, W. (1973) On fatigue analysis and testing for the design of the airframe, AGARD LS 62.
(42) SCHIJVE, J. (1985) The significance of flight-simulation fatigue tests, Technical University of Delft, Report LR-466.
(43) SCHÜTZ, W. (1972) The fatigue life under three different load spectra. Tests and calculations, AGARD-CP-118.
(44) HEULER, P. and SEEGER, T. (1986) A criterion for omission of variable amplitude loading histories, *Int. J. Fatigue*, **8**, 225–230.
(45) OPPERMANN, H. (1988) Zulässige Verkürzung zufallsartiger Lastfolgen für Betriebsfestigkeitsversuche, LBF-Koll, LBF-Bericht TB-180.

(46) AICHER, W., *et al.* (1976) Description of a fighter aircraft loading standard for fatigue evaluation 'FALSTAFF', Common Report of F + W Emmen, LBF, NLR, IABG.

(47) PERRET, B. H. E. (1985) TTCP collaborative programme on waveform analysis, Review of the work in the United Kingdom on the fatigue of aircraft structures 1983–1985.

(48) KRÜGER, W., SCHENTSOW, M., BESTE, A., and PETERSEN, J. Markov–Rainflow-Rekonstruktionen stochastischer Beanspruchungs-Zeit-Funktionen, *Fortschrittsberichte*, **18**.

(49) BERGMANN, J. W. (1983–1985) Rainflow analysis and synthesis, Review of investigations on aeronautical fatigue in the Federal Republic of Germany, LBF-Report, p. 173.

(50) HAIBACH, E. (1989) *Betriebsfestigkeit, Verfahren und Daten zur Bauteilberechnung*, VDI-Verlag Düsseldorf, Germany.

(51) (1980) Steel, concrete and composite bridges, BS 5400, Code of Practice for Fatigue, BS Institution, London, Part 10.

(52) Eurocode 3, design of steel structures, Part 1: general rules and rules for buildings. Provisional. Commision of the European Communities. EC3-88-C11-D4.

(53) HÜCK, M. and SCHÜTZ, W. (1978) Zum Stand der Lebensdauervorhersage – Beurteilung der Verfahran, IABG-Bericht TF-694.1, IABG-Bericht TF-694.2.

(54) HEULER, P. and SCHÜTZ, W. (1991) Lebensdauervorhersage für schwingbelastete Bauteile – Grundprobleme und Ansätze, *Aluminium*, **4–6**.

(55) SCHÜTZ, W. (1980) Lebensdauervorhersage schwingend beanspruchter Bauteile. Werkstoffermüdung und Bauteilfestigkeit – Prüfmethoden, Auswertung und Interpretation, DVM-Kolloquim Berlin.

(56) SCHÜTZ, W. (1983) Lebensdauer schwingend beanspruchter Bauteile, *Der Maschinenschaden*, **6**, 221–229.

(57) SCHÜTZ, W. (1985) Problematik der Lebensdauervorhersage für schwingend beanspruchte Bauteile (1–3), *Aluminium*, **2–4**.

(58) HEULER, P. and SCHÜTZ, W. Fatigue life prediction in the crack initiation and crack propagation stages, Proceedings of the 13th ICAF Symposium on aeronautical fatigue, (Edited by A. Salvetti and G. Cavallini), EMAS, Warley, UK.

(59) HEULER, P. and SCHÜTZ, W. (1991) Lebensdauervorhersage für schwingbelastete Bauteile – Grundprobleme und Ansätze (1–3), *Aluminium*, **4–6**.

(60) HÜCK, M., HEULER, P., and BERGMANN, J. (1991) Gemeinschaftsarbeit Pkw-Industrie/IABG, Relative Minerregel, Part II. Ergebnisse der Bauteilversuche. IABG-Bericht TF-2904.

(61) BUXBAUM, O., *et al.* (1983) Vergleich der Lebensdauervorhersage nach dem Kerbgrundkonzept und dem Nennspannungskonzept, LBF-Bericht Nr. FB-169.

M. A. Pompetzki* and T. H. Topper*

The Least and Most Damaging Histories Which Can Be Constructed From a Specific From–To Matrix

REFERENCE Pompetzki, M. A. and Topper, T. H. **The least and most damaging histories which can be constructed from a specific from–to matrix.** *Fatigue Design*, ESIS 16 (Edited by J. Solin, G. Marquis, A. Siljander, and S. Sipilä), 1993, Mechanical Engineering Publications, London, pp. 19–35.

ABSTRACT The From–To or Markov matrix is used as a compact storage technique to represent a service load history by storing the magnitude of excursions between successive reversals. Since the sequence of the excursions is not stored, a history reconstructed using the From–To matrix will have a different sequence of excursions than the measured service history (original history), giving rise to a different distribution of sizes of closed loops and a different calculated fatigue life. It is, therefore, of interest to determine the range of measured service histories that can result in the same From–To matrix. This paper presents a technique for estimating the most damaging and least damaging histories which can be constructed from a specific From–To matrix. These extreme histories are constructed according to damage hypotheses that either maximize or minimize the size of the Rainflow counted strain ranges and their mean stresses.

Introduction

The From–To matrix is a compact storage technique that provides a statistical representation of the measured strain history (1)–(3). This two-dimensional matrix stores the number of excursions from one strain level (From) to the next strain level (To) within the measured history, as shown in Fig. 1. The main diagonal represents excursions from one level to the same level, which do not contribute to fatigue damage and are undefined. The portion of the matrix above and to the right of the diagonal represents the number of excursions from a valley to a peak, while the lower left portion represents the number of excursions from a peak to a valley. A row in either the upper right or the lower left portion of the matrix is called a part row. The Markov matrix is easily constructed from the From–To matrix, by converting the number of excursions from one level to the next level, into the probability of an excursion occurring from one level to the next level (Fig. 1). Each part row in the Markov matrix must sum to 1.0. Other properties of the From–To matrix are explained by Haibach *et al.* (1).

The process of reconstructing the strain history from the From–To matrix involves randomly selecting a subsequent strain level given the current strain level and the transitional probabilities. The transitional probabilities for an excursion starting at a peak are given by the part row on the lower left portion of the Markov matrix, while a part row in the upper right portion of the matrix gives the transitional probabilities for an excursion starting at a valley.

* Department of Civil Engineering, University of Waterloo, Waterloo, Ontario, Canada.

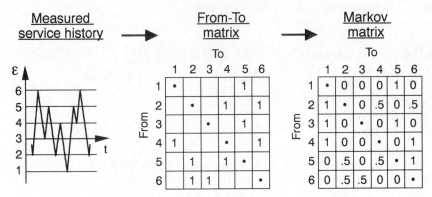

Fig 1   Procedure for creating the From–To and Markov matrices from measured service history

Since the history is reconstructed using randomly selected excursions, the sequence of excursions for the reconstructed history will not match those of the measured service history. As a result, the distribution of the sizes of the closed loops and calculated fatigue life will also differ from those of the measured service history.

When the service conditions are stored using the From–To matrix, it is desirable to know the range of measured service histories, based on calculated damage, that can result in a given matrix. This paper presents a technique for constructing the most damaging and least damaging histories for a given From–To matrix, thereby providing an estimate of the upper and lower bounds on calculated damage. Recall that for the From–To matrix the distribution of excursions is stored and not the distribution of closed loops. For a fixed distribution of closed loops, both Perrett **(4)** and Buxbaum *et al.* **(5)** found that the fatigue lives for various loading sequences fell within a factor of two on fatigue life. Since the From–To matrix does not contain sufficient information to reproduce the original distribution of closed loops for the measured service history, it is expected that the variation in calculated damage between the extreme histories will be larger for the From–To matrix than when the distribution of closed loops is stored.

Before describing the procedure for creating these two histories several new terms are defined, followed by a description of the damage parameters that form the criteria from which the upper and lower bounds are created.

Definitions

The *lowest level* is the lowest strain level in the From–To matrix that contains a reversal. Similarly, the *highest level* is the highest strain level that contains a reversal.

An *excursion* in strain is the movement in either the positive or negative direction from one reversal to the next reversal. A number of successive excursions is called a *sequence* of excursions. Under some conditions a sequence may contain a single excursion.

A *closed path* is a sequence of excursions that start and end with either a peak or a valley at the same strain level. The largest strain level in a closed path is the *maximum level* and the smallest strain level is the *minimum level*. A closed path describes one Rainflow counted closed loop between a maximum level and a minimum level and possibly several smaller closed loops (*internal cycles*). Closed paths can be divided into two portions: the negative portion, from the maximum level to the minimum level and the positive portion, from the minimum to the maximum level. Within each portion a peak or valley at any strain level cannot be repeated. Two examples of closed paths are shown in Fig. 2. Since the start and end point is given twice in the numerical representation of a closed path, but only occurs once in the actual closed path, the starting reversal is shown in brackets to indicate it is not part of the seqeuence, but forms part of the first excursion. Generally, the maximum level or the minimum level is used as the start and end point for the numerical representation of a closed path.

Damage parameters

The damage or change in crack length caused by a cycle in a strain history increases as the strain range of the cycle increases and as the mean stress of the cycle increases. The extreme histories either maximize or minimize the size of closed paths (strain range) and then maximize or minimize the mean stress of the closed paths. The effect of loading sequence, which can also have a significant influence on fatigue damage, is not considered here but will be dealt with in a future paper.

Strain range
Strain range is the single most important damage parameter for fatigue calculations and under variable amplitude loading, damage is based on the sum of all Rainflow counted strain ranges. Recall that the reconstructed history from

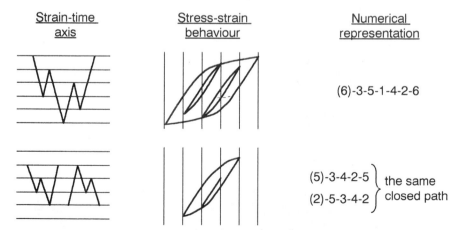

Strain-time axis	Stress-strain behaviour	Numerical representation
		(6)-3-5-1-4-2-6
		(5)-3-4-2-5 } the same
		(2)-5-3-4-2 } closed path

Fig 2   **Two examples of closed paths**

the From–To matrix is created by successively selecting excursions. Small excursions can be combined to create a Rainflow counted cycle, where the overall strain range is larger than any excursion taken individually. As shown in Fig. 3, the damage due to this sequence (group A) is significantly greater than the total damage for the situation when the excursions are combined to form several smaller cycles (group B). The damage calculation shown in Fig. 3 assumes that damage is proportional to strain range raised to a power of n, where n is taken as 3 in this example. The parameter, n, is the negative inverse of the slope of the strain–life curve, where the slope typically varies between -0.5 at short fatigue lives and -0.1 at long fatigue lives for metals.

The term *range content* is used as a measure of the total damage for a particular history, based solely on the size of the Rainflow counted strain ranges. Range content is calculated as the sum of all Rainflow counted strain ranges raised to the power n and is used to compare the Rainflow cycle count of different histories. Although the value of n can vary between 2 and 10, its magnitude is not crucial for determining range content, since range content is only used for comparisons and any value of n greater that 1 reveals the same trends as shown in Fig. 3. When comparing two histories with equal From–To matrices, a higher range content indicates that the history causes more damage, based on the magnitude of the Rainflow counted strain ranges.

Mean stress
For a closed path with a given maximum and minimum strain level, damage increases with mean stress. To maximize damage in terms of mean stress,

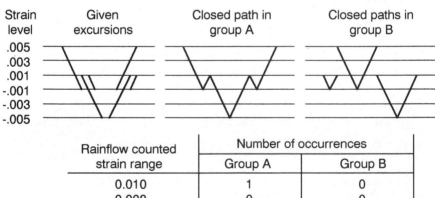

Strain level	Given excursions	Closed path in group A	Closed paths in group B

| Rainflow counted strain range | Number of occurrences | |
	Group A	Group B
0.010	1	0
0.008	0	0
0.006	0	2
0.004	0	0
0.002	2	1

Damage $\propto$ $\dfrac{\text{Range}}{\text{content}} = \Sigma \, \Delta\varepsilon^n$		
Assume $n=3$	1.008×10^{-6}	0.224×10^{-6}

Fig 3   From the same excursions, two groups of closed paths and their range content

closed paths are positioned on the positive portion of the outermost available loop, while, minimum damage is obtained by positioning closed paths on the negative portion of the outermost available loop.

Overview of the technique

The From–To matrix describes the magnitude and number of excursions in the measured service history. In order to maintain a given matrix when reconstructing a history of a given length, the number and size of excursions cannot be changed, only the magnitude of the Rainflow counted cycles and the sequence of excursions can be altered. An estimate for the upper bound on damage is constructed by first maximizing the number of occurrences of the largest overall strain range. For each occurrence this involves creating a closed path that starts at the maximum level, proceeds to the minimum, and returns to the maximum. Each closed path contains one cycle with a strain range equal to the maximum overall strain range and possibly several smaller cycles. Subsequently, the number of occurrences of the second largest strain range is maximized and this process is repeated for all strain ranges until each excursion in the From–To matrix has been used once. Closed paths that are created for strain ranges less than the maximum are positioned in such a way as to maximize their mean stress. The history estimating the lower bound is created by minimizing the number of occurrences of the largest available strain range and minimizing the mean stress of each closed path.

The histories representing the upper and lower bounds maximize and minimize damage, respectively, by maximizing or minimizing the range content and mean stress of the stress–strain loops. They are, however, only estimates of the true upper and lower bounds, because as well as sequence, a few situations have been omitted in this investigation. These situations, which have been shown to cause only small changes in total damage, do not appear to justify the effort required to include them in the construction of the extreme histories.

Upper bound on damage

To obtain an estimate of the most damaging history from a given From–To matrix two steps are used. First, the history is converted into a group of closed paths that maximize the range content and, second, each closed path is positioned in a location that maximizes its mean stress.

Maximize range content

The maximum range content is obtained by selecting closed paths that maximize the size of the Rainflow counted strain ranges. Recall that a closed path is a sequence of excursions starting at a maximum level, proceeding to a minimum level, and returning to the maximum level. To maximize range content the maximum number of closed paths are selected for each of the successively decreasing strain ranges, starting with the largest overall strain range. After all closed paths for the overall largest strain range have been found (at least one occurrence must exist), the other strain ranges are selected in decreasing order and the process is repeated. Within each strain range the number of closed paths for the highest mean strain level is maximized first, followed by maximization of the number of loops for each of the available mean strain levels in decreasing order. This results in the maximum possible mean stress of each closed path, since the higher the maximum strain for a given strain range, the higher the closed path will be positioned on the outer stress–strain loop.

The selection of a closed path is determined in two segments, first a sequence of excursions that start at the maximum level and end at the minimum level (negative portion) is selected and second, a sequence of excursions from the minimum level to the maximum (positive portion) is selected. The techniques for determining the excursions that make up each portion of a closed path are similar. A detailed description is given for selecting the negative portion of a closed path, while for the positive portion, only an outline of the technique is given, highlighting differences from the technique described for the negative portion.

Selecting the negative portion of a closed path
The negative portion of a closed path is determined by establishing a table of all non-repeating peaks and valleys, between the maximum and minimum levels, which can be attained from the maximum. When the minimum level is in the list, the negative portion of a closed path must exist.

The process for selecting the negative portion of a closed path is described with the aid of Fig. 4, where the maximum is level 6 and the minimum is level 1. A peak at the maximum level (level 6) is selected as the starting point. All

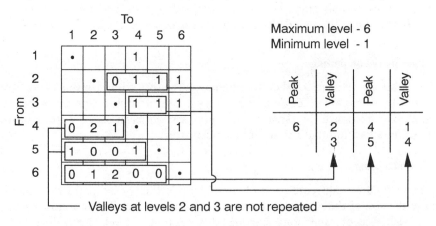

Negative portion of the closed path: 6-3-5-1

Fig 4 Selecting the negative portion of a closed path when maximizing range content

valleys (levels 2 and 3) that can be attained directly from this peak are given by the entries in part row 6 of the From–To matrix. Valleys at any stage in the process will be greater than or equal to the current minimum level, since excursions to levels below the minimum would already have been extracted from the matrix when creating larger excursions. In the present case the minimum level is not in the list of valleys (levels 2 and 3) and a list of peaks (levels 4 and 5) are established that can be attained from the previously selected valleys. Peaks at any stage in the process will be less than the maximum level, since any peak cannot be repeated within the negative portion of the closed path and excursions to higher levels would already have been extracted from the matrix. From the current list of peaks (levels 4 and 5) a new series of valleys are selected (levels 1 and 4), ignoring any valleys that have been previously used. For the situation in Fig. 4, the minimum level is now in the list of valleys and, therefore, a sequence of excursions exist from the maximum level to the minimum level. If the minimum level is not in the list, this process is repeated until either the minimum level is attained or the list of the current peaks or valleys is empty. If the current list is empty, the minimum level cannot be attained and, therefore, a closed path between the current maximum and minimum levels does not exist.

When the minimum is in the current list of valleys, the sequence of excursions that represents the negative portion of the closed path must be extracted from the table that contains the attainable peaks and valleys. The negative portion of the closed path is extracted by working backwards, starting with the valley at the minimum level, which is in the current list of valleys. Next the lowest attainable previous peak (level 5) is selected followed by the highest attainable previous valley (level 3) until the sequence of excursions is obtained. Remember that valleys must be lower than peaks and peaks higher than valleys. By selecting the lowest peaks and the highest valleys the size of internal cycles is kept to a minimum. Internal cycles are Rainflow counted cycles within the closed path that have a strain range smaller than the strain range of the closed path. When minimizing the size of internal cycles, small excursions are used for creating the desired closed paths. This leaves the medium size excursions available for creating closed paths and as a result range content is maximized (Fig. 5). After the negative portion of the closed path has been selected, the corresponding excursions are removed from the From–To matrix.

Selecting the positive portion of a closed path
When the negative portion of a closed path exists, the positive portion must also exist, given that the original From–To matrix is valid. A valid From–To matrix requires that the number of excursions arriving at a peak or valley equals the number of excursions that leave that peak or valley.

The same procedure as that just described for the negative portion, is used for selecting the positive portion of the closed path. A valley at the minimum level is selected as the starting point and all attainable peaks and valleys are

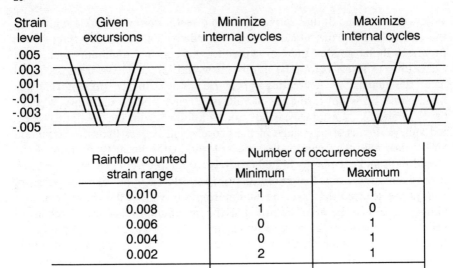

| Rainflow counted | Number of occurrences | |
strain range	Minimum	Maximum
0.010	1	1
0.008	1	0
0.006	0	1
0.004	0	1
0.002	2	1

Damage $\propto \dfrac{\text{Range}}{\text{content}} = \Sigma \, \Delta\varepsilon^n$

Assume $n=3$

| | 1.53×10^{-6} | 1.29×10^{-6} |

Fig 5   Maximize range content by minimizing the size of internal cycles

determined, until a peak at the maximum is attained. When this maximum level is found, the corresponding sequence of excursions is extracted from the table of valleys and peaks by working backwards from this maximum until the minimum level is obtained. The size of internal cycles are minimized by selecting the highest attainable valleys and lowest attainable peaks. After the positive portion of the closed path has been selected, the corresponding excursions are removed from the From–To matrix.

Constructing the history

Once the maximum range content has been obtained the closed paths are combined to create a load history that represents the upper bound for fatigue damage. The history is created by first establishing a basic sequence that contains at least one occurrence of each closed path positioned at its optimum location. Once the basic sequence has been established optimum locations exist for all required strain levels. The remaining closed paths are positioned, at the locations that correspond to the highest stress location available for their given maximum strain, within the basic sequence.

The procedure for constructing the basic sequence is complex since the closed paths must be combined so that the magnitude of the excursions are not altered. That is, the history constructed must result in a From–To matrix that is identical with the original From–To matrix. The following sections describe how a closed path is positioned, how the basic sequence is created, and finally how the remainder of the closed paths are positioned. The term partial basic sequence is used to describe the basic sequence during the process of its construction.

Positioning a closed path

The positioning of a closed path in the partial basic sequence is restricted because the closed path can only be positioned at reversals that are common to both the partial basic sequence and the closed path to be positioned. If a closed path is positioned where there is no corresponding reversal in the partial basic sequence, the magnitude of the excursions are altered and the resulting history will produce a From–To matrix that is different from the original From–To matrix. The example in Fig. 6 shows that the magnitude of the excursions have been altered by positioning closed path X in the partial basic sequence (Fig. 6(b)). The excursion from reversal 3b to 4 was not part of the original excursions in Fig. 6(a). Therefore, in order to position a closed path, it must contain a peak or valley at the same strain level as a peak or valley within the partial basic sequence. In Fig. 6(a), the closed path Y is positioned with its minimum at reversal 3 (Fig. 6(c)) and, subsequently, closed path X is positioned with its maximum at reversal 3a.

To obtain the upper bound on fatigue damage each closed path must be positioned so that its mean stress has the maximum possible value. This is done by positioning each closed path on the positive portion of the outermost loop available, as shown in Fig. 7. Therefore, to maximize mean stress it is necessary to position closed paths with their maximum at the highest available level. In general, when a closed path is positioned at its maximum or minimum level, the closed path does not affect the mean stress or range content of the remainder of the partial basic sequence.

Creating the basic sequence

This section describes the process of creating the basic sequence, which includes the technique for finding optimum locations from which closed paths

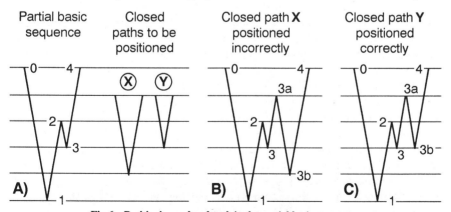

Fig 6   **Positioning a closed path in the partial basic sequence**

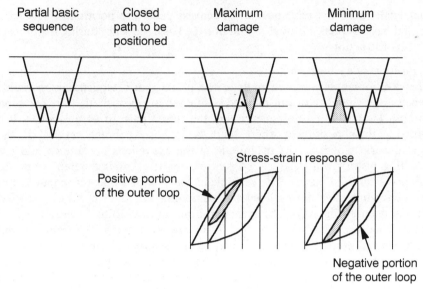

Fig 7   Positioning a closed path on the positive and negative portions of the outer most loop

are positioned, for creating optimum locations on the outermost stress–strain loop and for determining the next largest stress–strain path on which an optimum location may exist or can be created.

The basic sequence is a series of excursions that contains reversals at optimum locations such that all remaining closed paths can be positioned from their maximum level, thereby, maximizing their mean stress. If one occurrence of each closed path is included in the basic sequence then optimum locations for each maximum level will exist. The construction of the basic sequence is started by positioning all the closed paths that have a maximum level equal to the highest level. At least one closed path must exist between the highest and lowest levels. Subsequently, all closed paths with a minimum level equal to the lowest level, are positioned (Fig. 8). At this point the partial basic sequence will probably contain several optimum locations. These optimum locations are reversals (peaks, except at the lowest level) that occur on the positive portion of the outermost stress–strain loop, as shown in Fig. 8.

If optimum peaks exist at all strain levels that correspond to maxima of closed paths, then the basic sequence is complete and the remaining closed paths can be positioned, this is, however, unlikely. It is probable that optimum peaks do not exist within the partial basic sequence at several of these strain levels. The process of creating optimum peaks starts with the largest level at which an optimum peak is required (desired level). Before trying to create a reversal point at the desired level, one occurrence of each closed path that has a maximum above the desired level is positioned. To create an optimum location at the desired level a reversal must exist on the

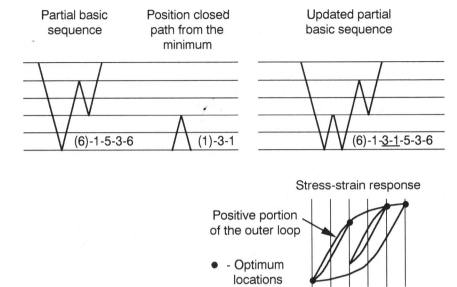

Fig 8   Optimum locations on the positive portion of the outermost loop

outermost loop below the desired level. If one or more reversals do exist below the desired level, a list of valleys and peaks is created by starting at the desired level in the partial basic sequence and working backwards until the lowest level is attained. Next, the positive portion of all closed paths with a maximum level equal to the desired level are checked for peaks or valleys that match a value within the list. If a match occurs that particular closed path is positioned from the matching reversal, thus creating an optimum peak in the partial basic sequence at the desired level (Fig. 6). If a match does not occur, the next highest strain level that requires an optimum peak is selected and the process is repeated.

When an optimum peak is not found or cannot be created on the positive portion of the outermost stress–strain loop, the stress–strain path representing the next largest positive path is selected and the process is repeated. If the next largest path occurs at a larger minimum strain than the previous path, all closed paths that have a minimum level equal to the minimum of the current outermost stress–strain path are positioned at that minimum level. Subsequently, the current outermost stress–strain path is examined for any reversal occurring at a strain level that corresponds to a required optimum peak. If any optimum peaks are still required, the previously described process is used, in an attempt to create a reversal at the desired strain level. This process is repeated until an optimum peak exists at all strain levels corresponding to the maximum level of every closed path. At this point the basic sequence is complete and the remaining closed paths can be positioned.

Positioning the remainder of the history

After the basic sequence is completed, optimum levels exist at each required strain level, such that the remaining closed paths can be positioned from their maximum levels. It is probable that the basic sequence contains more than one peak at each required optimum level and when a closed path is positioned from its maximum an additional peak is created at that optimum level. Since there may be many peaks in the sequence that occur at a given optimum location, the remaining closed paths are distributed uniformly over the peaks that correspond with this optimum location.

Lower bound on damage

Since the process for creating the least damaging history (lower bound on damage) parallels that for the most damaging history, only an outline of the process of creating the least damaging history is presented, highlighting differences from the most damaging history, already described. Two steps are used to create the lower bound on damage. First, the history is converted into a group of closed paths that minimize the range content and, second, each closed path is positioned in a location that minimizes its mean stress.

Minimizing range content

The minimum range content is obtained by selecting closed paths that minimize the size of the Rainflow counted strain ranges. A closed path, in this case, starts at the minimum level proceeds to the maximum and returns to the minimum, since closed paths are positioned from their minimum level in order to minimize mean stress.

At least one occurrence of the largest overall strain range must exist. The closed path is selected using the technique described above, but the size of the internal cycles within any closed path is maximized by selecting the highest peaks and the lowest valleys from the table of attainable peaks and valleys. This is the opposite of the procedure for obtaining the upper bound, in which the size of internal cycles was minimized in order to maximize range content. After the largest overall strain range has been determined, the remaining closed paths are selected starting with the largest single excursion with the lowest minimum stress. In Fig. 9, the largest excursion in the From–To matrix is the excursion from level 1 to level 5. These levels are then selected as the minimum and maximum levels and a closed path between these levels is determined as just described. As the positive and negative portions of a closed path are determined the corresponding excursions are removed from the From–To matrix.

Periodically a closed path does not exist within the limits defined by the largest single excursion. When this occurs, as in Fig. 10, the limits of the table of attainable peaks and valleys are expanded to the lowest and highest levels in the From–To matrix and the attainable peaks and valleys are again selected. The closed path is determined as shown in Fig. 10.

Largest single excursion with lowest minimum level. ➤ From: 5 To: 1
Find the positive portion of the closed path.

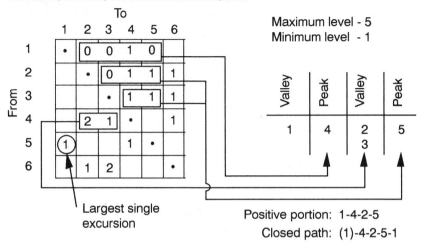

Fig 9 Selecting a closed path when minimizing range content

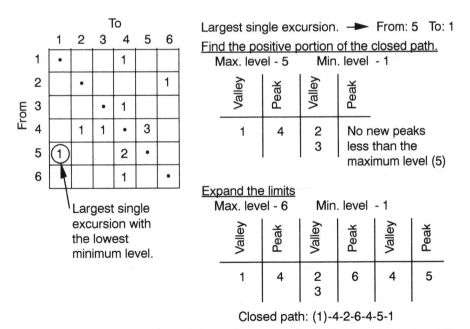

Fig 10 Selecting a closed path when one does not exist for the limits defined by the largest single excursion

Constructing the history

Closed paths are combined in order to create a history that represents the lower bound for fatigue damage. As with the upper bound, a basic sequence is established that contains reversals (valleys, except at the highest level) at optimum locations such that all remaining closed paths can be positioned from their minimum level, thereby minimizing their mean stress. The process of creating the history is the same as that described above, except that for the lower bound on damage closed paths are positioned from their minimum level on the negative portion of the outermost loop available. Once the basic sequence is complete the remaining closed paths are positioned with their minimum levels at optimum valleys.

Preliminary results

An analysis was performed using the standard history Gauss **(1)**, which is based on a stationary Gaussian process. Three versions of the Gauss history were used, each specified by a different irregularity factor (I), where the irregularity factor is the ratio of the number of zero level crossings with a positive slope to the number of peaks. The Gauss history was chosen since one version provides a From–To matrix that is representative of general fatigue applications ($I = 0.7$), while the other two matrices represent extremes ($I = 0.99$ and $I = 0.3$). A history constructed from a given From–To matrix is called a reconstructed history, since it represents a reconstruction of the original process. Reconstruction is done on a reversal-by-reversal basis, given the current reversal, the transitional probabilities and a pseudo random number generator. Generally, two reconstruction techniques are used: 'withdrawal with replacement', in which the transitional probabilities do not vary over time and 'withdrawal without replacement', in which the transitional probabilities are continuously adjusted to account for excursions which have already occurred. The subsequent results include a reconstructed history, which is taken as the average result of three reconstructed histories using 'without replacement' for each version of the Gauss history. An average result provides an estimate for any reconstructed history that would be obtained from a given From–To matrix using either with or without replacement, since both techniques give similar average values.

The spectrum shape is shown in Fig. 11 for the reconstructed Gauss histories together with the upper and lower bounds on damage. The spectrum for each upper bound and the lower bound for $I = 0.99$ is not significantly different from that of the reconstructed history, but the lower bounds for $I = 0.7$ and $I = 0.3$ are significantly different from the reconstructed history. Theoretically, a measured service history with the same From–To matrix could have a spectrum shape anywhere between the upper and lower bounds. The large difference between the reconstructed history and the history representing the

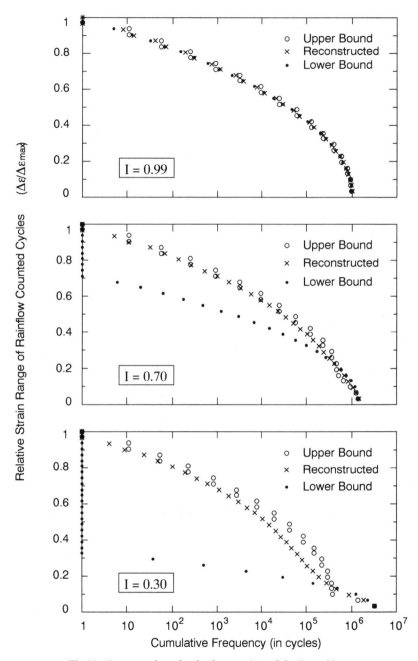

Fig 11   Spectrum shape for the three versions of the Gauss history

Table 1   Calculated damage with and without mean stress correction on the Gauss history

History → Gauss	Total damage	
$\Delta\varepsilon_{max} = 0.007$ (I-irregularity factor)	Without mean stress correction	With mean stress correction
I = 0.99 Upper bound	0.399 70	0.462 400
Reconstructed	0.394 30	0.394 800
Lower bound	0.383 60	0.304 600
I = 0.7 Upper bound	0.378 20	0.597 900
Reconstructed	0.284 70	0.295 700
Lower bound	0.120 80	0.045 480
I = 0.3 Upper bound	0.261 80	0.405 100
Reconstructed	0.112 50	0.120 500
Lower bound	0.005 82	0.000 920

lower bound will be examined, using other measured service histories, in a subsequent investigation.

A damage analysis was performed using an SAE 1045 steel, where the strain–life curve under variable amplitude loading is represented as a straight line that extends below the constant amplitude fatigue limit. This strain–life representation takes into account the damage done in variable amplitude loading by small cycles below the constant amplitude fatigue limit (6)–(8). For the SAE 1045 steel, the slope of the line in terms of strain amplitude and reversals to failure is -0.23 and the intercept is 0.04 (8)(9). Total damage is shown in Table 1, for the three versions of the Gauss history and the histories representing the upper and lower bound. Damage is calculated using both Rainflow counted strain ranges (without a mean stress correction) and the Smith–Watson–Topper parameter (10) to account for mean stresses.

Conclusions

A technique is presented for estimating the most damaging and least damaging histories which can be constructed for a specific From–To matrix. The most damaging history is constructed by creating closed paths that maximize the size of the Rainflow counted cycles and then positioning them so that their mean stress is maximized. The least damaging history is constructed by creating closed paths that minimize the size of the Rainflow counted strain ranges and then positioning them so that their mean stress is minimized. The calculated damage values for the three versions of the Gauss history show that the history representing the lower bound can be significantly different than the reconstructed history or the upper bound.

Acknowledgements

Financial support of this research by the National Sciences and Engineering Research Council of Canada and GKN Technology Incorporated is gratefully acknowledged.

References

(1) HAIBACH, E., FISCHER, R., SCHÜTZ, W., and HÜCK, M. (1976) A standard random load sequence of gaussian type recommended for general application in fatigue testing; its mathematical background and digital generation, *Fatigue testing and design*, (Edited by R. G. Bathgate), Vol. 2, SEE, UK, pp. 29.1–29.21.

(2) HÜCK, M. and SCHÜTZ, W. (1975) Generating the FALSTAFF load history by digital mini computers, Problems with fatigue in aircraft, Proceedings of the eighth ICAF symposium, (Edited by J. Branger, and F. Berger), pp. 3.62/1–3.62/23.

(3) CONLE, F. A. and TOPPER, T. H. (1982) Fatigue service histories: techniques for data collection and history reconstruction, *SAE Technical Paper No. 820093*. Society of Automotive Engineers.

(4) PERRETT, B. H. E. (1987) An evaluation of a method of reconstituting fatigue loading from Rainflow counting, *New materials and fatigue resistant aircraft design, Proceeding of the fourteenth ICAF Symposium*. Engineering Materials Advisory Services Ltd., pp. 355–401.

(5) BUXBAUM, O., KLÄRSCHKE, H., and OPPERMANN, H. (1991) Effect of loading sequence on the fatigue life of notched specimens made from steel and aluminium alloys, *Appl. Mechs Revs*, **44**, 27–35.

(6) HEULER, P. and SEEGER, T. (1986) A criterion for omission of variable amplitude loading histories, *Int. J. Fatigue*, **8**, 225–230.

(7) POMPETZKI, M. A., TOPPER, T. H., and DuQUESNAY, D. L. (1990) Effect of compressive underloads and tensile overloads on fatigue damage accumulation in SAE 1045 steel, *Int J. Fatigue*, **12**, 207–213.

(8) DuQUESNAY, D. L. (1991) *Fatigue damage accumulation in metals subjected to high mean stress and overload cycles*. PhD Thesis, University of Waterloo, Canada.

(9) DOWDELL, D. J., LEIPHOLZ, H. H. E., and TOPPER, T. H. (1986) The modified life law applied to SAE-1045 steel, *Int. J. Fracture*, **31**, 29–36.

(10) SMITH, K. N., WATSON, P., and TOPPER, T. H. (1970) A stress–strain function for the fatigue of metals, *J. Mats*, **5**, 767–778.

G. Hénaff,* J. Petit,* and N. Ranganathan*

Damage Tolerance of a Helicopter Rotor High-Strength Steel

REFERENCE Hénaff, G., Petit, J., and Ranganathan, N., **Damage tolerance of a helicopter rotor high-strength steel,** *Fatigue Design,* ESIS 16 (Edited by J. Solin, G. Marquis, A. Siljander, and S. Sipilä) 1993, Mechanical Engineering Publications, London, pp. 37–49.

ABSTRACT The near-threshold fatigue crack growth of a high strength low alloy steel under variable amplitude loading is investigated in ambient air and in vacuum. Although strong interaction effects are brought out, no concomitant closure is detected due to the high mean level of the different loading blocks. A model of the observed retardation phenomenon, which incorporates crack growth laws previously developed to describe the influence of environment, is then proposed and shows good agreement with experimental data.

Introduction

In 1989 international flight regulation authorities imposed the requirement to design helicopter rotors according to the damage tolerance concept (FAR 29.571). Although this concept has been successfully used for many years for aeroplanes, its extension to helicopters necessitates the need for further investigation of the fatigue crack growth behaviour of the materials used in such structures.

A helicopter rotor piece endures about 20 000 loading cycles per hour, which induces a total life time containing 10^2 times more fatigue cycles than for an aeroplane. From the point of view of fatigue crack propagation, one must no longer confine the analysis to the so-called Paris regime, but also consider the slow crack growth rate range, i.e., the near-threshold regime. In this regime, environmental influence, which is of prime interest in the case of helicopters, is known to provide a significant contribution to fatigue crack propagation. Moreover, the fatigue design of these structures must take into account the variable aspect of in-service loadings and, nowadays, only very few data are available on the behaviour under loading spectra of high-strength steels used in helicopter rotors, especially in this crack growth rate range.

The present study comes within this framework. The near-threshold fatigue crack growth behaviour of the 30NCD16 steel of the French nomenclature is investigated. Constant amplitude tests are first conducted in order to obtain the reference behaviour. Program blocks are then derived from actual spectra. They constitute an experimental tool for the study of interaction effects which may exist between different load levels. All of these tests are conducted in air and in vacuum in order to gain insight into the environmental contribution to the observed phenomena.

* Laboratoire de Mécanique et de Physique des Matériaux, URA CNRS 863, ENSMA, 20 rue Guillaume VII, 86034 Poitiers Cedex, France.

Experimental conditions

The material is a high-strength low-alloy steel 30NCD16 in a quenched and tempered state. Chemical composition and mechanical properties are given in Tables 1 and 2 respectively.

This material is more particularly used in designing the so-called 'baby's bottle' piece shown in Fig. 1.

Test specimens were compact–tension type with the following dimensions: $W = 75$ mm and $B = 12$ mm.

Fatigue tests were conducted at 35 Hz on a servo-hydraulic machine equipped with a chamber (1) specially designed to provide controlled atmospheres or high vacuum conditions (2×10^{-3} Pa). The crack length was optically measured by means of a travelling microscope. For variable amplitude testing, a micro-computer was used to monitor the machine.

Crack closure measurements were performed during each experiment at test frequency using the compliance technique and a digital oscilloscope.

The study of the behaviour under loading spectra has been conducted by considering the following load sequences:

Table 1   Chemical composition of 30NCD16 steel

Element	C	Si	Mn	Ni	Cr	Mo
% weight	0.298	0.36	0.45	3.44	1.30	0.44

Table 2   Mechanical properties of 30NCD16 steel

E (GPa)	R_e (MPa)	R_m (MPa)	A %
191	1130	1270	13

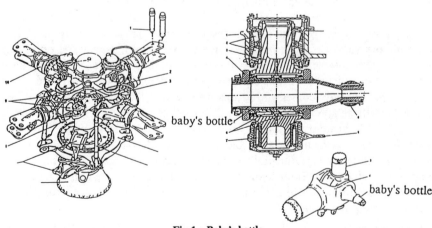

baby's bottle

baby's bottle

Fig 1   Baby's bottle

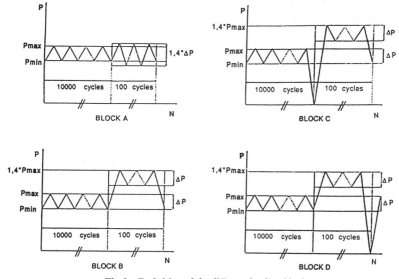

Fig 2   **Definition of the different loading blocks**

- *repeated ground-air-ground cycles:* a zero load is applied every 1000 cycles of the baseline loading ($R = 0.7$);
- *constant mean level and variable amplitude:* the amplitude is increased by 40 percent without any change on the mean level during 100 cycles every 10^4 cycles of the baseline level (block A on Fig. 2);
- *constant amplitude and variable mean level:* the maximum load is increased by 40 percent while the amplitude remains unchanged during 100 cycles every 10^4 cycles of the baseline level (block B on Fig. 2).

The blocks in Fig. 2 were designed so as to bring to the fore the eventual sequence effects. Indeed, in each block, only one parameter varied and the crack advance during the high level cycles could be neglected when estimated according to a cumulative rule.

After preliminary experiments on block B, it was thought that the introduction of a zero load, owing to experimental procedure, may disturb crack growth rates. It was, therefore, decided to study the effects of the addition of a ground–air–ground (GAG) cycle onto block B, either just before the high level cycles (block C in Fig. 2) or just after these cycles (block D in Fig. 2).

Results

Constant-amplitude reference testing

Fatigue crack propagation on the 30NCD16 steel in ambient air and in vacuum have been extensively discussed elsewhere (2)(4) and crack growth enhancement in ambient air has been particularly emphasized; the following conclusions have been reached:

(a) The intrinsic propagation, i.e., after crack closure correction and without any environmental influence as in vacuum, can be described, in the range $(10^{-10}$ m/cycle $\leqslant da/dN \leqslant 10^{-7}$ m/cycle), by the following relationship **(2)(4)**

$$\frac{da}{dN} = \frac{A}{D_0^*} \left(\frac{\Delta K_{eff}}{G} \right)^4 \tag{1}$$

where ΔK_{eff} denotes the effective value of the cyclic stress intensity factor as defined by Elber **(5)**, G the shear modulus, and D_0^* the critical value of the cumulative displacement at the crack tip leading to crack advance.

(b) The effective propagation in air can be interpreted as the superposition of two distinct processes: the adsorption of water vapour molecules on freshly created surfaces which does fundamentally not affect the implied mechanism but promotes crack growth by reducing the value of D_0^* down to D_{ad}^*, and a propagation mode assisted by hydrogen resulting from the dissociation of the adsorbed water vapour molecules and then dragged ahead of the crack tip by moving dislocations **(3)**. A crack growth law reflecting this superposition has been proposed

$$\frac{da}{dN} = \frac{A}{D_{ad}^*} \left(\frac{\Delta K_{eff}}{G} \right)^4 + B \frac{\Delta K_{eff}^2 - \Delta K_{eff, th}^2}{ER_e} \tag{2}$$

Figure 3 compares fatigue data in air and computation of the proposed law. Good agreement is found over the whole explored range.

Variable amplitude tests

Repeated ground–air–ground cycles
It can be seen in Fig. 4 that this kind of spectrum does not disturb the baseline loading $(R = 0.7)$ in air.

Block A
Figure 5 presents the results obtained in air. No effect is noticed on the measured crack growth rates.

Block B
The results obtained for this block programme in air and vacuum are shown in Figs 6 and 7, respectively. A slowing down of the growth rates can be observed in both environments although the general trend is reversed. Thus, in air, this retardation effect reaches its maximum near threshold where the crack growth rates are divided by a ratio of about 25, and begins to decrease as the value of the applied ΔK increases. In vacuum, however, this effect is hardly noticed near threshold and only appears with increasing ΔK. Another salient

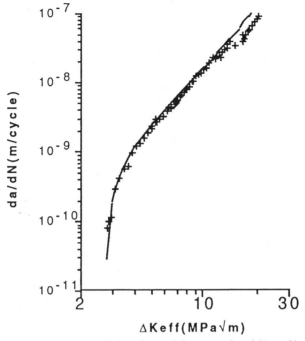

Fig 3   Comparison between fatigue data and the proposed model in ambient air

feature of this slowing down effect is that no concomitant crack closure was detected during the experiments.

Blocks C and D
These two types of loading have also been studied both in air and vacuum and the results are shown in Figs 6 and 7, respectively.

It can be seen that in the two environments block C behaves almost like block B. But for block D, a single application of a GAG cycle just after the overloads completely annuls the retardation effect, so that the measured crack growth rates in this case are equal to the rates that would be obtained under a baseline level loading of the same amplitude. Hence one has to consider two characteristic behaviours: (a) one without interaction effect, i.e., similar to constant amplitude loading at $R = 0.7$ (blocks A and D); and (b) one characterized by a slowed down propagation (blocks B and C). The following aims to analyse this retarded propagation by assuming that this phenomenon results from the classical delay observed under 'high–low' type loading.

Analysis

A special experimental procedure was developed in order to finely monitor the

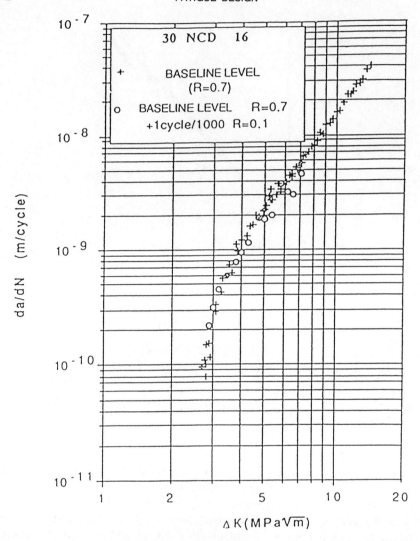

Fig 4   Results obtained under repeated GAG cycles in air

crack advance just after the application of the 100 high-level cycles. Two different kinds of test were carried out:

(a) a single block applied under manual control: after the 100 high-level cycles, the load is decreased down to its baseline level;
(b) programme blocks similar to block B but with various values of N_t, the number of baseline cycles between overloads.

In the second case, it is assumed that the crack advance is the same for each

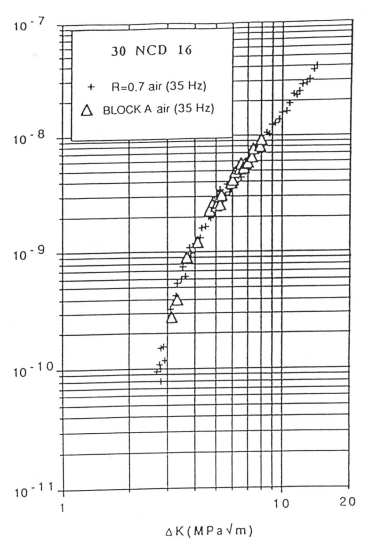

Fig 5   Results obtained under block A in air

block. This has been verified experimentally. The crack advance during one block Δa is then deduced from the total advance measured after the application of several blocks by dividing this advance by the number of blocks. By this method, one can obtain typical retardation curves as presented in Fig. 8 for $\Delta K = 10$ MPa$\sqrt{m}$ in air and in vacuum. These curves have similar trends at other values of ΔK (6). The similarity between the two curves on Fig. 8 suggests that environment does not affect the retardation process, as previously noticed by Ranganathan et al. (7).

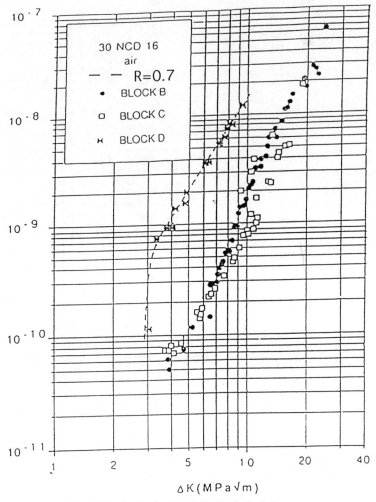

Fig 6   Results obtained in air for blocks B, C, D

Four characteristic stages can be distinguished on these curves (see Fig. 9):

(a) during the first stage the crack behaves as if no overloads have been applied (no retardation);

(b) the second stage is characterized by an incremental crack growth (retardation);

(c) the propagation is accelerated throughout the third stage ('lost retardation' **(8)**)

(d) in the fourth stage the crack has recovered its nominal growth rate (steady crack growth).

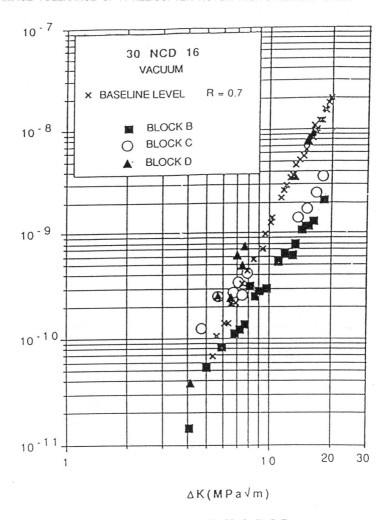

Fig 7   Results obtained in vacuum for blocks B, C, D

It has been shown that these curves provide an explanation for the different behaviour between air and vacuum under block B conditions by considering the stage of the crack development after 10^4 cycles (6). For example, in ambient air, near threshold, the crack tip is in the second stage where retardation reaches its maximum, but as ΔK increases, it enters into the third stage where retardation decreases; this is why the slowing down effect on growth rates becomes less pronounced. At threshold in vacuum, however, the crack tip just reaches the second stage, so that retardation and consequently the slowing down effect are quite negligible. As the ΔK value increases, the crack

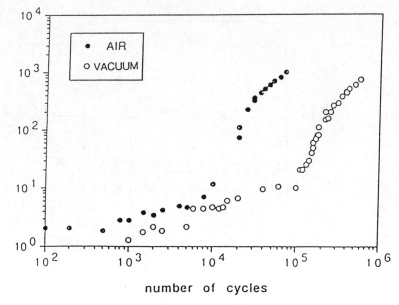

Fig 8   Retardation curves for $\Delta K = 10\mathrm{MPa}\sqrt{\mathrm{m}}$ in air and vacuum

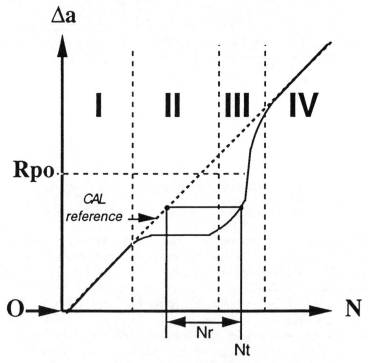

Fig 9   Definition of the different stages on retardation curves

grows deeper into the second stage, and rate at which the crack growth rate slows down increases.

In order to take into account differences in growth rates that exist under constant amplitude loading conditions between various ΔK levels and environments, the following parameters have been introduced

$$-x = \frac{\Delta a}{R_{po}}$$

which denotes the rate of progression of the crack tip in the overload's plastic zone (R_{Po}: size of the overloads plastic zone);

$$-\frac{N_r}{N_t}$$

where N_r is the number of delay cycles and N_t the total number of applied cycles (see Fig. 9).

Figure 10 presents all retardation data using these parameters, for different values of ΔK and for the two environments. It can be seen that all these points fall onto a single curve called the retardation mastercurve. This mastercurve may be considered as a representation of the retardation mechanism within the perturbed zone. Retardation in this case is assumed to be due to residual stresses since no crack closure is detected. Furthermore, crack closure would not account for the 'lost retardation' phenomenon. An analytical expression of this mastercurve has been derived

$$\frac{N_r}{Nt} = A \exp\left\{-\alpha\left|\ln\left(\frac{x}{x_{max}}\right)\right|^n\right\} \tag{3}$$

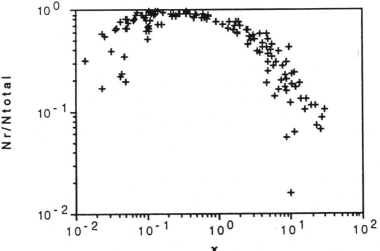

Fig 10   Retardation data in the perturbed zone in both environments

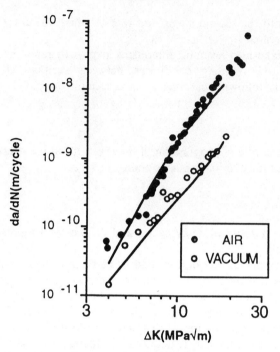

Fig 11   Comparison between predicted and measured crack growth rates in the case of block B

where:

A = maximum of the mastercurve ($A = (N_r/N_t)_{max}$);

x_{max} = point of maximum retardation;

α, n = parameters describing the evolution of retardation or the gradient of residual stresses, that must be fitted to experimental data.

This equation takes into account the lack of retardation either just after the application of the overloads ($x \ll x_{max}$), or far away from the point of overload application ($x \gg x_{max}$) (see Figure 9). Crack growth rates under blocks are then easily computed by using the fatigue crack propagation law suitable for the considered environment (see above) for a given ΔK value. In Fig. 11 the predictions obtained in the case of block B are compared to experimental values. Good agreement is observed in both environments. Therefore, a procedure for the evaluation of materials under this type of spectra can be proposed. It consists of the determination of the retardation mastercurve (Fig. 10) which provides the parameters needed to compute crack growth rates by means of equation (3). This would thus allow for a considerable reduction in the number of time-consuming tests.

Conclusion

The fatigue crack growth behaviour of a high-strength steel has been investigated in air and in vacuum under block loading representative of helicopter spectra. The near-threshold regime was considered with a high baseline level ($R = 0.7$). Strong sequence effects have been brought out, in particular:

(a) an important slowing down of crack growth after 100 cycles with an increase of 40 percent on maximum load while keeping the amplitude constant;

(b) a complete vanishing of this effect by adding a GAG cycle just after the overload cycles (block D).

These effects take place without any concomitant crack closure.

The observed slowing-down effect on crack growth rates has been correlated to the retardation induced by post-overload interaction.

An analytical formula has been derived to describe this retarded behaviour.

A special procedure, based upon the determination of a retardation mastercurve has been proposed to determine the parameters needed to compute the formula. Ongoing experiments will aim to check the extensibility of this approach to other spectra and materials.

Acknowledgements

This work was carried out under the financial support of Aérospatiale Joint Research Center in Suresnes under the BE-3386-89 BRITE/EURAM "DAMTOL" project.

The authors also wish to thank B. Journet for his help and valuable suggestions.

References

(1) PETIT, J., NADEAU, A., LAFARIE, M. C. and RANGANATHAN, N. (1980) Réalisation d'un caisson hermétique pour essais dynamiques, *Revue de Physique Appliquée*, **15**, 919, in French.

(2) PETIT, J. and HÉNAFF, G. (1991) Intrinsic stage II fatigue crack propagation, *Scripta Met. Materiala*, 2683.

(3) HÉNAFF, G., BOUCHET, B. and PETIT, J. (1992) Environmental influence on the near-threshold fatigue crack propagation behaviour of a high strength steel, *Int. J. Fatigue*, **14**, 211.

(4) HÉNAFF, G. and PETIT, J. (1992) An analysis of the environmental influence on the near-threshold behaviour of a high-strength low-alloy steel, *Proceedings of ECF9*, (Edited by S. Sedmak, A. Sedmak, and D. Ruzic), EMAS, pp. 433–438.

(5) ELBER, W. (1971) The significance of crack closure, *Damage tolerance in aircrafts structures*, *ASTM STP 486*, ASTM Philadelphia, p. 230.

(6) HÉNAFF, G., PETIT, J., and JOURNET, B. (1992) Fatigue crack propagation behaviour under variable amplitude loading in the near-threshold region of a high-strength low-alloy steel, *Fatigue Fracture Engng Mater. Structures*, **15**, 1155.

(7) RANGANATHAN, N., QUINTARD, M., PETIT, J. and DE FOUQUET, J. (1990) Environmental influence of the effect of a single overload on the fatigue crack growth behaviour of a high strength aluminium alloy, *Environmentally Assisted Cracking Science and Engineering*, *ASTM STP 1049*, (Edited by W. B. Lisagor, T. W. Crooker and N. B. Lies) ASTM, Philadelphia, p. 374.

(8) BERNARD, P. J., LINDLEY, T. C., and RICHARDS, C. E. (1972) Mechanics of overload retardation during fatigue crack propagation, *Fatigue Crack Growth under Spectrum Loads*, *ASTM STP 595*, ASTM Philadelphia, p. 78.

A. Vašek* and J. Polák*

Fatigue Life of Two Steels under Variable Amplitude Loading

REFERENCE Vašek, A. and Polák, J., **Fatigue life of two steels under variable amplitude loading**, *Fatigue Design*, ESIS 16 (Edited by J. Solin, G. Marquis, A. Siljander, and S. Sipilä) 1993, Mechanical Engineering Publications, London, pp. 51–61.

ABSTRACT Smooth cylindrical specimens made from a very low carbon steel and a martensitic cast steel were subjected to various loading histories arranged as a sequence of blocks with variable strain amplitudes. The number of cycles to failure was measured as a function of the maximum amplitude in a block. The effect of the ratio of low to high amplitudes in a block was studied. Fatigue lives depend strongly on both the maximum amplitude in a block and the relative frequency of the lowest and highest amplitudes in a block.

Several fatigue life prediction methods were adopted. The best agreement with experiments was obtained using the methods based on the behaviour of short cracks in the materials studied.

Introduction

The non-deterministic feature of service loading as a sequence of variable amplitudes causes many difficulties and uncertainties in fatigue life evaluations. A knowledge of the fatigue behaviour of materials in constant amplitude loading is usually insufficient for reliable life prediction (1). A correct analysis of a loading history and a realistic damage accumulation rule are prerequisites to achieve this complicated task.

The Rainflow counting method is the most appropriate method of analysis of a variable amplitude loading history (2). Very often the loading history can be divided into sections or blocks that are repeated periodically until fracture. Then statistical parameters of a loading history can be substituted by parameters of one loading block.

A reliable fatigue life prediction method requires a realistic rule for fatigue damage accumulation. The simplest and quite successful damage accumulation rule is Miner's rule (3). Among the many modifications of Miner's rule, the approach of Corten and Dolan (4), and later that of Kliman (5) have proved to be most successful. The linear damage rule has not given a good agreement with experimental data (6)–(8), and, therefore, the double linear damage rule has been proposed (7). In recent years an investigation of crack initiation and propagation in constant and variable amplitude loading (9)–(14) has led to an improvement in fatigue life prediction methods.

In the present paper fatigue life curves of two steels under two-step block loading and variable amplitude block loading are measured. The effect of the maximum amplitude in a block and the effect of the ratio of low to high amplitudes in a block were studied. Experimental data are compared with

* Academy of Sciences of the Czech Republic, Institute of Physical Metallurgy, Brno, Czech Republic.

Table 1   Chemical analysis of the tested steels (in wt%)

Steel	C	Mn	Si	P	S	Cr	Ni	Cu	Mo
Low carbon steel	0.008	0.2	—	0.007	0.02	0.06	0.07	0.03	0.02
Martensitic cast steel	0.4	0.72	0.33	0.021	0.023	13.1	6.4	0.12	0.51

fatigue lives estimated by means of several prediction methods including the method based on the behaviour of short fatigue cracks (see Appendix).

Experiments

Cylindrical specimens having a 10 mm diameter and a 15 mm gauge length were made from two steels. Very low carbon steel (0.008%C) was annealed at 600°C for 1 h in vacuum. Average grain size was 40 μm. A large block of martensitic cast steel (13%Cr, 6%Ni, 0.5%Mo) was annealed at 990°C for 12 h, cooled, and subsequently tempered for 15 h at 590°C and slowly cooled to room temperature. Average prior austenitic grain size was 80 μm and residual austenite content was 9 percent. The chemical composition of both steels is shown in Table 1.

The specimens were subjected to two-step block loading and to random amplitude block loading in an electro-hydraulic computer-controlled testing machine (MTS 880). Total strain was controlled with the sine wave and frequency, resulting in a constant average strain rate of $5 \times 10^{-3} \text{ s}^{-1}$. The stress and strain peaks were continuously monitored and stored on the disk. Rainflow analysis of loading history was applied to obtain full cycles and amplitudes for later fatigue life calculations.

The onset of a macroscopic crack resulted in a characteristic change in the hysteresis loop shape and in a decrease in stress amplitude. When the stress amplitude of the largest loop in a block dropped to 80 percent of the saturated value, the specimen was concluded to have failed. The total number of applied cycles until failure determined fatigue life.

Low carbon steel was subjected to: (a) two-step block loading (Fig. 1) with varying relative frequency of low and high amplitudes in a block; and (b) to random amplitude block loading with varying distribution of loading amplitudes (A, B, and C in Fig. 1) and fixed interval of amplitudes. Martensitic cast steel was subjected to: (a) two-step block loading (Fig. 1); and (b) to random amplitude block loading with fixed amplitude distribution (D in Fig. 1) and varying maximum amplitude in a block.

Results

Low carbon steel

The two-step loading block consisted of n_H cycles with high strain amplitude ($\varepsilon_{aH} = 6 \times 10^{-3}$) and n_L cycles with low strain amplitude ($\varepsilon_{aL} = 1 \times 10^{-3}$). n_H

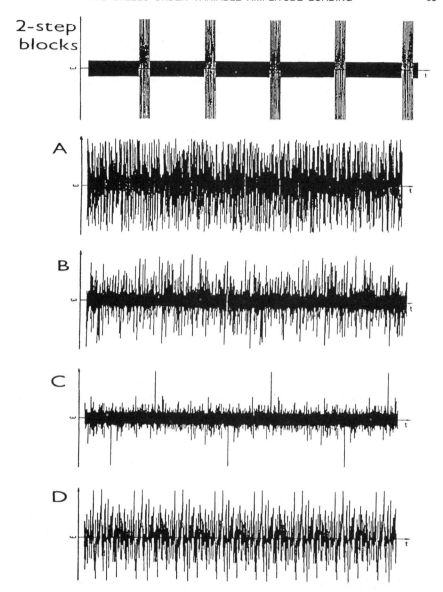

Fig 1 **Strain versus time loading histories: two-step block loading, three types of random block loading with power-law amplitude distribution, A, B, and C, and random block loading with exponential amplitude distribution (D)**

was held constant at 10, and the ratio n_L/n_H was varied from 1 to 3000. The fatigue life in constant amplitude cycling with ε_{aH} was $N_H = 3150$ cycles to failure, and with ε_{aL} was $N_L = 260\,000$ cycles to failure. The ● in Fig. 2 show fatigue lives of low carbon steel under two-step block loading. In Fig. 2 fatigue

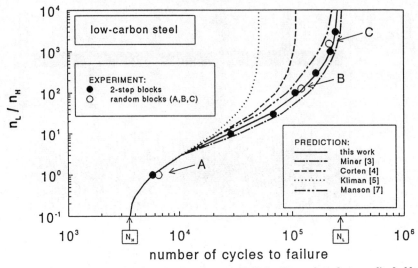

Fig 2 Fatigue life of low carbon steel under two-step block loading and random amplitude block loading with varying ratio n_L/n_H (types A, B, and C) and constant value of the maximum and minimum amplitude in a block

lives are expressed by the total number of cycles to failure while varying ratio n_L/n_H.

Three types of random amplitude block loading with three distributions A, B, and C of strain amplitudes and constant width of loading spectrum were applied to the low carbon steel specimens. The probability density function of the amplitude ε_a was

$$f_{A, B, C}(\varepsilon_a) = \frac{\alpha + 1}{\varepsilon_{aH}} \frac{\left(\dfrac{\varepsilon_a}{\varepsilon_{aH}}\right)^\alpha}{1 - \left(\dfrac{\varepsilon_{aL}}{\varepsilon_{aH}}\right)^{\alpha+1}}, \quad \text{for } \varepsilon_{aL} \leq \varepsilon_a \leq \varepsilon_{aH} \tag{1}$$

$\varepsilon_{aL} = 1 \times 10^{-3}$ was the lowest amplitude and $\varepsilon_{aH} = 6 \times 10^{-3}$ was the highest amplitude in a block. The exponents for these three distributions were $\alpha_A = 0$, $\alpha_B = -2.7$, and $\alpha_C = -4.1$. One block contained 500 random amplitudes. Ratio n_L/n_H of the number of cycles having the lowest amplitude ε_{aL} and the highest amplitude ε_{aH} in a block can be derived from equation (1)

$$\frac{n_L}{n_H} = \frac{f(\varepsilon_{aL})}{f(\varepsilon_{aH})} = \left(\frac{\varepsilon_{aL}}{\varepsilon_{aH}}\right)^\alpha \tag{2}$$

Therefore, the ratio n_L/n_H may characterize the loading history with power-law distribution of amplitudes (1) and the experimental fatigue lives obtained in random amplitude block loading with amplitude distribution (1) could be plotted in the same diagram together with the two-step block loading life data (see ○ in Fig. 2). Both data sets can be fitted by a single curve.

Martensitic cast steel

The two-step loading block consisted of $n_H = 10$ cycles of high strain amplitude $\varepsilon_{aH} = 4.6 \times 10^{-3}$ and n_L cycles of low strain amplitude $\varepsilon_{aL} = 2 \times 10^{-3}$. The ratio n_L/n_H was varied from 0.3 to 3000. The fatigue life in the constant amplitude cycling with ε_{aH} was $N_H = 3500$ cycles to failure, and with ε_{aL} was $N_L = 223\,000$ cycles to failure. Figure 3 shows experimental lives which lie on the curve having the same shape as the curve in Fig. 2.

Amplitudes in the random loading block were exponentially distributed with the probability density function

$$f_D(\varepsilon_a) = \frac{1}{\varepsilon_{am}} \exp\left(-\frac{\varepsilon_a}{\varepsilon_{am}}\right) \tag{3}$$

where ε_{am} is the mean value of strain amplitudes. Loading blocks having 250 amplitudes with different maximum amplitude ε_{aH} in the interval from 2×10^{-3} to 1.9×10^{-2} were generated. Amplitudes ε_a below $\varepsilon_{aL} = 1.5 \times 10^{-4}$ were cut off. Distribution (3) was designated as type D. Figure 4 shows experimental fatigue lives versus the maximum amplitude in a block.

Fatigue life prediction

Five fatigue life prediction methods (9)(3)–(5)(7) were adopted to estimate the fatigue resistance of low carbon steel and martensitic cast steel under variable amplitude loading. A brief description of fatigue life calculations is given in the

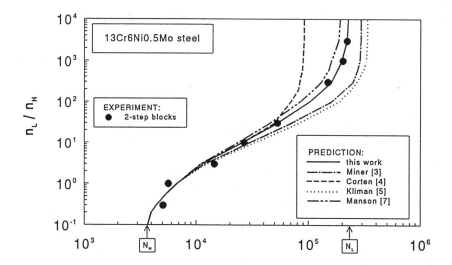

number of cycles to failure

Fig 3 Fatigue life of martensitic cast steel under two-step loading with varying ratio n_L/n_H and constant value of the maximum and minimum amplitude in a block

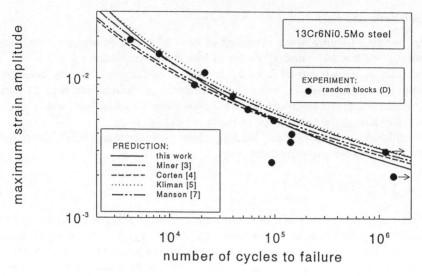

Fig 4   Fatigue life of martensitic cast steel under random amplitude block loading (type D) with varying maximum amplitude in a block

Appendix. Predicted fatigue life curves are drawn in Figs 2–4 together with experimental data.

Discussion

Figure 2 shows the fatigue life curve of low carbon steel under block loading with a constant width of amplitude spectrum ($\varepsilon_{aL} < \varepsilon_a < \varepsilon_{aH}$) and a varying ratio of low to high amplitudes in a block n_L/n_H. The number of cycles to failure depends strongly on the ratio n_L/n_H for values within the range (1–1000). Both high and low level amplitudes in a block affect the fatigue life in this interval. Outside this range, the number of cycles to failure does not depend on the ratio n_L/n_H and approaches constant amplitude fatigue life at high (N_H) or low (N_L) amplitude (see N_H, N_L in Figs 2 and 3). For $n_L/n_H < 1$, high amplitudes have a dominant effect and low amplitudes can be neglected; for $n_L/n_H > 1000$, however, low amplitudes have a dominant role and the presence of high amplitudes is not significant for fatigue life.

Similar results can be derived from Fig. 3. The knees in the fatigue life curve appear for $n_L/n_H = 1$ and $n_L/n_H = 1000$. An identical interval of the ratio n_L/n_H was found both with low carbon steel and with martensitic cast steel. This is mainly due to almost identical fatigue lives N_L, N_H for the lowest and the highest strain amplitude in a block for both steels though the strain amplitudes differ considerably. Therefore, the knees in the fatigue life curves in Figs 2 and 3 are given by the fatigue lives N_L and N_H at low and high strain amplitude. More exactly, the fatigue life depends on the fatigue damage caused by high and low amplitudes, which are approximately inversely proportional to the fatigue lives. Figure 4 shows that the fatigue life of the martensitic cast steel under random amplitude block loading with exponential distribution of strain amplitudes is strongly dependent on the maximum strain amplitude in a block within the whole range of tests performed. A significant scatter of experimental data close to fatigue limit was observed. Similar results were obtained recently in the study of low carbon steels (16) and Cr–Mo steel (6). Premature fracture of some specimens was identified with the presence of primary macroscopic defects on the specimen surface.

Several curves calculated using different prediction procedures are presented in Figs 2–4. The fatigue life prediction method based on the study of the short fatigue crack behaviour (see Appendix), gives the best agreement with experimental data. A good agreement in the whole interval of tests and for both steels was obtained using the Manson's method (7) and also using the simpler Miner's method (3). The Kliman's method (5) gives a non-conservative prediction for martensitic cast steel and a conservative prediction for low carbon steel. The prediction calculated by the Corten's method (4) is conservative for both steels.

Conclusions

Fatigue life measurements of low carbon steel and martensitic cast steel under variable amplitude block loading and subsequent analysis allow us to draw several conclusions.

(1) The fatigue life of the steels under loading with constant width of amplitude spectrum is strongly dependent on the ratio n_L/n_H. The interval was found for both steels (1, 1000). Outside this interval, fatigue life is independent of n_L/n_H and is determined only by the constant amplitude fatigue life corresponding to the highest amplitude in a block (when $n_L/n_H < 1$) and by the constant amplitude fatigue life corresponding to the lowest amplitude in a block (when $n_L/n_H > 1000$).

(2) The fatigue life of the steels under loading with exponential distribution of amplitudes strongly depends on the maximum strain amplitude in a block.

(3) The fatigue life prediction method based on the knowledge of the behaviour of short cracks yields the best agreement with experimental data. A good agreement was also obtained by the Manson's method and by the simple Miner's method.

Appendix – fatigue life prediction methods

Fatigue life evaluation formulae are shown for the case of a repeated loading block consisting n cycles with random amplitudes. Individual loading amplitudes $\{\varepsilon_{ai}\}_{i=1,...,n}$ in a loading block are characterized by constant amplitude fatigue lives $\{N_i\}_{i=1,...,n}$ on these levels. The fatigue lives at the maximum and minimum loading amplitude are designated as N_H and N_L respectively. The fatigue life at variable amplitude block loading is expressed as a total number of individual cycles to failure N_f.

Miner's method **(3)**

$$N_f = n\left[\sum_i^n \frac{1}{N_i}\right]^{-1} \tag{4}$$

Corten's method **(4)**

$$N_f = n\left[\sum_i^n \frac{1}{N_H}\left(\frac{N_H}{N_i}\right)^{0.8}\right]^{-1} \tag{5}$$

Kliman's method **(5)**

$$N_f = n\left[\sum_i^n \frac{1}{N_H}\left(\frac{N_H}{N_i}\right)^{c(m+1)}\right]^{-1} \tag{6}$$

where m is an exponent of cyclic stress-strain curve and c is an exponent of Manson–Coffin fatigue-life relation **(1)**

Manson's method **(7)**

$$N_f = n\left(\left[\sum_i^n \frac{1}{N_{Ii}}\right]^{-1} + \left[\sum_i^n \frac{1}{N_{IIi}}\right]^{-1}\right) \tag{7}$$

where

$$N_{Ii} = N_i \exp\left(\Omega N_i^\Phi\right)$$
$$N_{IIi} = N_i - N_{Ii} \tag{8}$$

and

$$\Phi = \frac{1}{\ln\left(\dfrac{N_H}{N_L}\right)} \ln\left(\frac{\ln\left(0.35\left(\dfrac{N_H}{N_L}\right)^{0.25}\right)}{\ln\left(1 - 0.65\left(\dfrac{N_H}{N_L}\right)^{0.25}\right)}\right) \tag{9}$$

$$\Omega = \frac{\ln\left(0.35\left(\dfrac{N_H}{N_L}\right)^{0.25}\right)}{N_H^\Phi}$$

Short crack behaviour method

The fatigue life prediction method is based on damage accumulation during fatigue loading. Recent investigations of surface fatigue cracks in low carbon steel (9) and in martensitic cast steel (17) have led to a recognition of two stages in fatigue life with qualitatively different regimes in damage accumulation. Fatigue damage was defined as a relative crack length

$$D = \frac{a}{a_f} \tag{10}$$

where a_f is crack length at the moment of specimen failure.

During the early, crack initiation phase (in), an increment in fatigue damage per one cycle was found constant and equal to

$$\Delta D_{in} = \frac{1}{a_f} v_{in} \tag{11}$$

where $v_{in} = (da/dn)_{in}$ is crack growth rate in the crack initiation stage.

During the latter, crack propagation phase (pr), a damage increment per one cycle was found proportional to crack length a

$$\Delta D_{pr} = \frac{1}{a_f} \{v_{in} + k(a - a_{tr})\} \tag{12}$$

where k is a parameter which was evaluated for individual loading amplitudes solving equation

$$\frac{1}{k} \ln \left\{ \frac{k(a_f - a_{tr})}{v_{in}} + 1 \right\} - \frac{N}{2} = 0 \tag{13}$$

a_{tr} is the crack length at the crack initiation–crack propagation transition and was estimated by the expression

$$a_{tr} = A(\sigma_a - \sigma_0) \tag{14}$$

the validity of which was verified for some materials in (17). In equation (14), σ_a is the stress amplitude; parameters A and σ_0 are shown in Table 2.

Surface fatigue cracks in specimens of low carbon steel (9) and martensitic cast steel (17) reached the transition value, a_{tr}, at the number of cycles corresponding to the half of fatigue life $N/2$. Crack growth rate v_{in} in equation (11)

Table 2 Parameters of the relation (14) for some materials

Material	σ_o (MPa)	A ($mMPa^{-1}$)
Poly-crystalline copper	47	3.2×10^{-6}
Low carbon steel	78	5.8×10^{-7}
Martensitic cast steel	230	3.5×10^{-7}

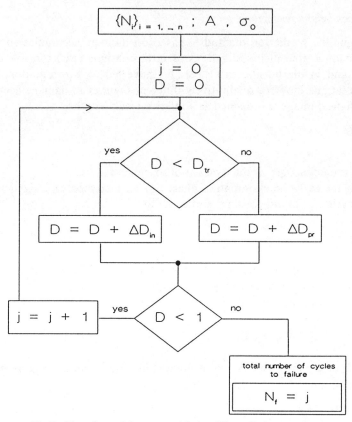

Fig 5   Flow chart of the presented fatigue life prediction (see above)

was then evaluated by the expression

$$v_{in} = \left(\frac{da}{dn}\right)_{in} = \frac{a_{tr}}{N/2} \tag{15}$$

Finally, fatigue damage accumulation can be described by the following:

$$D = D + \Delta D_{in}; \quad \text{for } D < D_{tr} = \frac{a_{tr}}{a_f}$$

$$D = D + \Delta D_{pr}; \quad \text{for } D \geqslant D_{tr} = \frac{a_{tr}}{a_f} \tag{16}$$

where fatigue damage, D, increases until reaching maximum value of damage, $D = 1$, characterizing specimen failure. The number of increments ΔD, including $D = 1$, defines the number of cycles to failure N_f

$$N_f = \sum_{D=0}^{D=1} j \tag{17}$$

A flow chart of this fatigue life prediction method is shown in Fig. 5.

References

(1) POLÁK, J. (1991) Cyclic plasticity and low cycle fatigue life of metals, Elsevier, Amsterdam.

(2) *Fatigue damage measurement and evaluation under complex loadings*, (1991), (Edited by Y. Murakami).

(3) MINER, M. A. (1945) Cumulative damage in fatigue, *J. appl. Mech.*, **12**, A159.

(4) CORTEN, T. H. and DOLAN, T. J. (1956) Cumulative fatigue damage, *Proceedings of International Conference on Fatigue of Metals*, Institute of Mechanical Engineering, London, p. 235.

(5) KLIMAN, V. (1984) Fatigue life prediction for a material under programmable loading using the cyclic stress–strain properties, *Mater. Sci. Engng*, **68**, 1–10.

(6) POLÁK, J., VAŠEK, A., and KLESNIL, M. (1988) Effect of elevated temperatures on the low cycle fatigue of 2.25Cr-1Mo steel – Part II: variable amplitude straining. Solomon, H. D. et al. (eds). *Low cycle fatigue, ASTM STP 942*, (Edited by H. D. Solomon), ASTM, Philadelphia, pp. 922–937.

(7) MANSON, S. S. and HALFORD, G. R. (1986) Re-examination of cumulative fatigue damage analysis – an engineering perspective, *Engng Fracture Mech.*, **25**, 539–571.

(8) MILLER, K. J. and IBRAHIM, M. F. E. (1981) Damage accumulation during initiation and short crack growth regimes, *Fatigue Engng Mater. Structures*, **4**, 263–277.

(9) VAŠEK, A. and POLÁK, J. (1991) Low cycle fatigue damage accumulation in armco-iron, *Fatigue Fracture Engng Mater. Structures*, **14**, 193–204.

(10) SIEGEL, J., SCHIJVE, J., and PADMADINATO, U. H. (1991) Fractografic observations and predictions on fatigue crack growth in an aluminium alloy under mini TWIST flight-simulation loading, *Int. J. Fracture*, **13**, 139–147.

(11) LANKFORD, J. and HUDAK, Jr., S. J. (1988) Relevance of the small crack problems to lifetime predictions in gas turbines, *Int. J. Fatigue*, **9**, 87–93.

(12) HICKS, M. A. and PICKARD, A. C. (1988) Life prediction in turbine engines and the role of small cracks, *Mater. Sci. Engng* 43–48.

(13) EBI, C., RIEDEL, H., and NEUMAN, P. (1987) Fatigue life prediction based on microcrack growth, *Fracture control of engineering structures* (Edited by H. C. van Elst and A. Bakker) ECF6, pp. 1587–1593.

(14) HOSHIDE, T., MIYHARA, M., and INOHUE, T. (1987) Life prediction based on analysis of, crack coalescence in low cycle fatigue, *Engng Fracture Mech.*, **27**, 91–101.

(15) POLÁK, J. and VAŠEK, A. (1991) Analysis of the cyclic stress-strain response in variable amplitude loading using rainflow method, *The Rainflow method in fatigue*, (Edited by Y. Murakami), Butterworth, Oxford, pp. 123–131.

(16) POLÁK, J. and KLESNIL, M. (1979) Cyclic plasticity and low cycle fatigue life in variable amplitude loading, *Fatigue Engng Mater Structures*, **1** 123–133.

(17) POLÁK, J. et al. (1990) Fatigue life prediction based on the knowledge of short fatigue crack behaviour, Report VZ 811/950, Institute of Physical Metallurgy, Czechoslovak Academy of Sciences, Brno.

T. Dahle* and B. Larsson*

Spectrum Fatigue Data in Comparison to Design Curves in the Long Life Regime

REFERENCE   Dahle, T. and Larsson, B., **Spectrum fatigue data in comparison to design curves in the long life regime,** *Fatigue Design*, ESIS 16 (Edited by J. Solin, G. Marquis, A. Siljander, and S. Sipilä) 1993, Mechanical Engineering Publications, London, pp. 63–71.

ABSTRACT   Spectrum fatigue tests on welded steel box beam specimens were run at stress levels giving high cycle fatigue lives ($> 10^7$ cycles). A comparison of the test results with current fatigue design codes shows that the fatigue data are unsafe by a factor of approximately four. This was also shown by using a simple fracture mechanics prediction model. Therefore, new design principles and other models of crack growth in the long life regime are needed.

Notation

a	Crack length
$\Delta\sigma_{eq}$	The equivalent stress range
$\Delta\sigma_0$	The maximum stress range in a spectrum
ΔK_{eq}	Equivalent stress intensity factor range
γ_i	The frequency of each stress level of $\Delta\sigma_i$ of cycles n_i relative to total number of cycles in the design spectrum
θ_i	Ratio of $\Delta\sigma_i/\Delta\sigma_0$
K_f	Fatigue notch factor
K_t	Stress concentration factor
M_k	Global geometric correction factor in a growing crack field
N	Number of cycles
N_f	Number of cycles to fracture in the experiment
N_p	Predicted number of cycles to fracture

Introduction

Interest in high-cycle, long-life fatigue of welded components subjected to spectrum loading from offshore, transportation, and ship building industries (1)–(3) is increasing. A design life of 10^8 cycles or more is often required. However, very few investigations in this area are reported in the open literature. The reason is presumably non-technical, i.e. the testing is extremely time consuming and expensive. Nevertheless, the need for test data and theoretical analysis is apparent.

The object of this study is to report results from spectrum fatigue testing on beam specimens performed some years ago and to compare the results with current design codes.

* ABB Corporate Research, Laboratory for Applied Mechanics, S–721 78 Västerås, Sweden.

Table 1   Chemical composition and strength properties

Chemical composition

C	Mn	Si	P	V	Nb
(max wt%)					
0.20	1.6	0.5	0.035	0.15	0.015

Strength properties

Yield strength, R_{eL}	410 MPa
Ultimate tensile strength, R_m	551 MPa
Elongation, A_5	32%
Area reduction	39%

Material and test specimens

The test beams were made of a weldable micro-alloyed steel according to the Swedish standard SS 2134–01 ($\approx$ ISO Steel Grade E 355) with chemical composition and mechanical properties according to Table 1.

Twelve half-scale by section and full thickness box beams (Fig. 1) were manufactured in three local workshops. The beams were manually arc-welded with longitudinal half 'V' groove welds at the bottom flange and double fillet welds at the top flange to a quality corresponding to WB-class ($\approx$ quality level C according to ISO (4)). All twelve beams were stress relieved at 580°C in an industrial furnace followed by slow cooling.

Test procedures

In-field strain measurements were conducted on a prototype railway bogie of fuel sized beams to obtain knowledge about the loading environment. The load history was characterized by the Rainflow-counting technique to establish a load spectrum.

The constructed test spectrum was truncated at the maximum level where the nominal bending stress was kept below the yield strength of the base material. The levels used in the tests are illustrated in Fig. 2 and are also shown in Table 2.

In order to shorten the time of testing, omission of the low amplitude load cycles is often performed. In our case, the possibility of such omission was

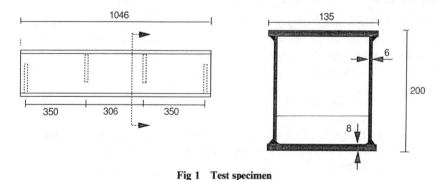

Fig 1   Test specimen

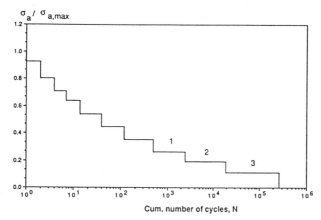

Fig 2 The test spectrum

investigated by testing small notched specimens with approximately the same estimated stress concentration as the beam welds. These tests were carried out with the same test spectrum as the beam specimens, but the lowest level or levels were omitted (omission levels 1, 2, 3 according to Fig. 2). The results indicated that even the lowest stress level (level 3) contributed to fatigue damage, even if this stress level was only about 25 percent of the constant amplitude fatigue limit or 11 percent of maximum stress amplitude in the spectrum. Thus, the conclusion of these pre-tests was that no omission of low stress cycles was permitted in the beam tests.

The beams were tested in a closed loop, electro-hydraulic computer-controlled test machine. The computer was fed with the load spectrum in a table from which a 'drawing without replacement' algorithm created a random load sequence on the specimen until the table was empty. This 'block' of loading consisted of 268 200 cycles and had an irregularity factor of close to one. A new load sequence was drawn within the next block and so on until fracture occurred or the test was stopped. Fracture was defined as 10 mm deflection of the mid-beam section.

Table 2 Test spectrum

i	$\sigma_{ai}/\sigma_{a,\,max}$	N	ΣN
1	1.000	1	1
2	0.922	1	2
3	0.800	2	4
4	0.707	3	7
5	0.634	7	14
6	0.537	26	40
7	0.444	80	120
8	0.346	380	500
9	0.259	2000	2500
10	0.190	16 800	19 300
11	0.109	248 900	268 200

Table 3   Test results. Stresses at the web/flange transition

Beam no.	σ_{max} MPa	$\Delta\sigma_{eq}$ MPa	Blocks to fracture	Cycles to fracture $\times 10^6$	Comment
VA2	368	44.9	194	52.2	
VA4	368	44.9	>362	>97.3	Stopped
VA5	368	44.9	250	67.2	
VA12	368	44.9	232	62.4	
VA3	322	39.3	201	54.1	
VA6	322	39.3	336	90.4	
VA7	322	39.3	>665	>179	Stopped
VA8	322	39.3	397	107	
VA1	315	65.9	684	13.2	Omitted
CA9	226	152	—	1.40	
CA10	226	152	—	1.70	
CA11	226	152	—	2.50	

Specimen no. VA1 was tested with the lowest stress level omitted.

The tests were carried out as four-point bending tests with constant mean stress at a test frequency of about 20 Hz. Four specimens were tested at a maximum stress of 368 MPa with mean stress 184 MPa and four specimens at 322 MPa with mean stress of 161 MPa, both with the spectrum as described in Fig. 2. One beam specimen was tested with the lowest stress level omitted yielding a block size of 19 300 cycles.

Three beam specimens were also tested at constant stress range of 152 MPa at a mean stress of 76 MPa.

Test results

The test results are summarized in Table 3 for both spectrum tests (VA-) and constant amplitude tests (CA-). As expected, the crack initiation site was located mostly on the rear side of the half 'V' groove weld at the bottom flange.

Discussion of results

Comparison to codes

Fatigue data for box beams are scarcely reported in available literature. Therefore, the test results from Table 3 are plotted in Fig. 3 together with the mean and 95 percent confidence limits of full-scale, manually welded 'I' beams fatigue data compiled and analyzed by Gurney and Maddox (8). The data are replotted using the equivalent stress range as proposed by Sahli and Albrecht (7) versus number of cycles to fracture. The equivalent stress can be written as

$$\Delta\sigma_{eq} = \left(\sum_i \gamma_i \theta_i^m \right)^{1/m} \times \Delta\sigma_0 \qquad (1)$$

where:

m is the slope of the S–N curve and is set = 3.0;
$\Delta\sigma_0$ spectrum maximum stress range;

γ_i the ratio between number of cycles at stress level $\Delta\sigma_i$, N_i, and total number of cycles in the spectrum, N_0;

θ_i ratio of $\Delta\sigma_i/\Delta\sigma_0$.

Using this approach it must be emphasized that the following conditions have to be met.

(a) No sequence effects are present or negligible.

(b) The linear damage rule is applicable within the complete life region considered (i.e., $< 10^8$ cycles).

(c) The S–N curve must be a straight line in the current life regime and expressed as $NS^m = \text{const}$. Subsequently, even small stress amplitudes below the constant amplitude fatigue limit will be damaging.

From an engineering point of view it is believed that the railway load spectrum represents a quasi-stationary random process with no extreme fluctuations of the maximum load. Furthermore, due to the dead weight, the R value will be high. This would result in no closure effects in crack propagation and subsequently no sequence effects.

Both constant and variable amplitude test results are plotted in Fig. 3. All tests lie within the scatter band of Gurney and Maddox's data **(8)** (represented by extrapolated curves for mean and 95 confidence limits).

According to Swedish design code **(5)** (fatigue strength curves are similar to the Eurocode **(6)**), the S–N curves to be used in the linear damage calculations are estimated with a change in slope at 5×10^6 cycles where $m_2 = 2m - 1$ and a 'cut-off-limit' at 10^8 cycles. All cycles with stress ranges below this 'cut-off-

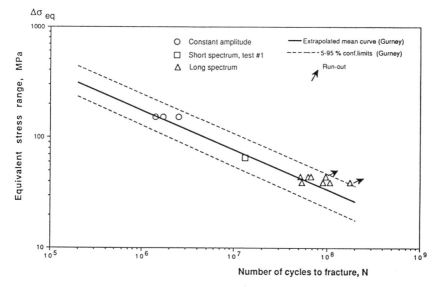

Fig 3   Results

Table 4   The life ratio

Spec no	$\Delta\sigma_{eq}$	N_f/N_p mean
VA2	44.9	0.22
VA5	44.9	0.28
VA12	44.9	0.26
VA3	39.3	0.12
VA6	39.3	0.21
VA8	39.3	0.25
VA1	65.9	0.38
Average:		0.25

limit' are ignored. The life ratio N_f/N_p (experiment divided by predicted life), based on the mean life curve in Fig. 3 corrected with the change in slope at 5×10^6 cycles and the cut-off-limit at 10^8 cycles, was calculated for the test results.

As can be seen in Table 4, the predicted life is unconservative by a factor of approximately four.

Our results agree with data from US investigators (2)(9)(10) of full-scale, spectrum loaded, beam tests. It was concluded from those studies that the existence of a fatigue limit below which no fatigue crack propagation occurs, is assured only if none of the stress cycles in the spectrum exceed the constant amplitude fatigue limit.

The constant stress range fatigue limit is estimated at 5×16^6 cycles to ≈ 100 MPa (5). This implies that for the tests run in this investigation about 99 percent of the load cycling was performed below the constant amplitude fatigue limit. The lowest stress range run in the tests was 35 MPa. Still, it is impossible to determine if the observed small tendency for the results to fall above the mean life is due to a threshold effect or other effects. Clearly, more data and analysis in this life regime is needed.

Prediction by a fracture mechanics model

Prediction of the fatigue lifes of the tested beams was performed using a simple fracture mechanics, crack growth model as demonstrated by Samuelsson (11) (suggested by Sahli and Albrecht (7)). Crack growth is modelled using Paris' law

$$\frac{da}{dN} = C\Delta K_{eq}^n \tag{2}$$

where

$$\Delta K_{eq} = Y\Delta\sigma_{eq}\sqrt{(\pi a)}$$

in which $\Delta\sigma_{eq}$ is calculated with constant slope of the S–N curve with $m = 3$, and including all cycles with $\Delta\sigma_i \geqslant \Delta\sigma_{th}$, where

$$\Delta\sigma_{th} = \Delta K_{th}/Y\sqrt{(\pi a)}$$

The geometry factor Y is determined as

$$Y = \frac{M_k M_e M_w}{\sqrt{Q}}$$

where:

M_k is the 'geometric correction factor' which accounts for the fact that the growing crack lies in a (global) geometric stress concentration;

$M_e M_w$ are the correction factors in accordance with Newman (12) where M_e accounts for the thickness effect, back and front face effect, plasticity effect and M_w represents the finite width correction;

$\sqrt{Q}$ is the elliptical shape factor of the crack (12).

Applying the above procedure implies that $\Delta K_{eq} = 0$ (and $da/dN = 0$) only if all stress levels in the load spectrum fall below the threshold stress, $\Delta\sigma_{th}$.

The unknown parameters of the above expressions are the M_k factor and the initial crack depth, a_i. These factors are estimated in the analysis which follows.

The fatigue notch factor as reported by Park and Lawrence (13) is written as

$$K_f = 1 + \frac{K_t - 1}{1 + a/r} \tag{3}$$

where K_t is the stress concentration factor, 'a' is the Peterson's material parameter and r is the notch root radius. Using Park and Lawrence (13) 'the worst-case notch' $a = r$ and setting $K_f = 2.0$ (which corresponds to weld category $C = 90$ according to (5)) gives $K_t = 3$. At the surface ($a/t \approx 0.001$) $M_k(a/t) = K_t$. Further, we put $M_k(a/t) = 1$ at $a/t = 0.1$. Smith and Gurney (14) proposed that M_k can be approximated by the equation

$$M_k(a/t) = D\left(\frac{a}{t}\right)^q \tag{4}$$

Inserting the above boundary values for M_k, the empirical constants q and D can be estimated to -0.24 and 0.58, respectively. By integrating equation (2), solving N, and comparing to the constant amplitude fatigue results, the initial crack depth, a_i, could be calibrated to ≈ 0.05 mm. A survey of available literature indicated that the mean initial crack size is a value between 0.05–0.12 mm so the calibrated value seems reasonable and is used in the predictions below. Since the tests were run with constant mean stress, a weighted mean value of the R ratio can be estimated to ≈ 0.7.

Equation (2) is numerically integrated setting $\Delta K_{th} = 3.1$ MPa$(m)^{1/2}$, $C = 1.3 \cdot 10^{-11}$ and $n = 2.73$ based on own experiments. As shown the predictions give unconservative results by a factor of ≈ 5 in life. The results of the predictions are shown in Table 5.

It can thus be concluded that the simple model described above does not apply very well in the long life regime. From 'a short crack problem' point of

Table 5 Life predictions

Spec no	$\Delta\sigma_{eq}$	N_f/N_p $\Delta K_{th} = 3.1$	N_f/N_p $\Delta K_{th} = 2.0$	N_f/N_p $\Delta K_{th} = 1.3$
VA2	44.9	0.17	0.41	0.96
VA5	44.9	0.21	0.53	1.18
VA12	44.9	0.20	0.50	1.14
VA3	39.3	0.09	0.16	0.53
VA6	39.3	0.15	0.38	0.89
VA8	39.3	0.34	0.45	1.05
Average:		0.19	0.41	0.96

view the use of an initial crack depth of 0.05 mm may be doubtful, but the lack of a persistent model in the long life regime might be a reasonable motivation for the engineering approach used here. Still, it is tempting to infer that the reason for the discrepancy is a 'short crack' problem. As seen in Table 5, simulating the 'short crack effect' by lowering the (long) crack threshold, ΔK_{th}, in the above calculations, leads to a much better prediction. This may be justified by the fact that 'short cracks' give lower crack thresholds than 'long cracks' as shown by Newman et al. (15).

Conclusion

The main conclusion which can be drawn from this investigation is that using the Palmgren–Miner cumulative damage rule with current design codes can give excessive stresses when the design life typically exceeds 10^7 cycles. For safe design the subsequent recommendation must be to use a straight line S–N relation extended below the constant amplitude fatigue limit.

Acknowledgment

ABB Traction AB is gratefully acknowledged for giving us permission to publish the results given herein.

References

(1) ALMAR-NÆSS, A. (1985) *Fatigue handbook* NTH, Trondheim, Norway.
(2) KEATING, P. B. and FISHER, J. W. (1989) High-cycle, Long-life fatigue behavior of welded steel details, *J. Construct. Steel Research* 12, 253–259.
(3) MUNSE, W. H., WILBUR, T. W., TELLALIAN, M. L., NICOLL, K., and WILSON, K. (1983) Fatigue characterization of fabricated ship details for design, Ship Structure Committee SSC–318, AD–A1403338.
(4) Draft International Standard ISO/DIS 5817.3 (1990) Arc-welded joints in steel – guidance on quality levels for imperfections 118.
(5) *Swedish regulations for steel structures BSK*, (1989), Publication 118. Swedish Institute of Steel Construction.
(6) Eurocode 3 (1984) Common unified code of practice for steel structures Industrial Processes, Building and Civil Engineering, Commission of the European Communities, EUR 8849.
(7) SAHLI. A. and ALBRECHT, P. (1984) Fatigue life of welded stiffeners with known initial cracks, *ASTM STP 833*, ASTM, Philadelphia, pp. 193–217.

(8) GURNEY, T. R. and MADDOX, S. J. (1972) *A re-analysis of fatigue data for welded joints in steel*, The Welding Institute, UK.

(9) KEATING, P. B. and FISHER, J. W. (1987) Fatigue behavior of variable loaded bridge details near the fatigue limit TRR 1118, TRR National Research Council, pp. 56–63.

(10) FISHER, J. W., MERTZ, D. R., and ZHONG, A. (1983) Steel bridge members under variable amplitude long life fatigue loading (1983) NCHRP Report 267, National Cooperative Highway Research Program.

(11) SAMUELSSON, J. (1988) *Fatigue design of vehicle components: methodology and application, Dissertation*, The Royal Institute of Technology, Stockholm.

(12) NEWMAN, J. C. (1979) A review and assessment of the stress-intensity factors for surface cracks, *ASTM STP 687*, ASTM Philadelphia, pp. 16–42.

(13) PARK, S. K. and LAWRENCE, F. V. (1989) Monte Carlo simulation of weldment fatigue strength (1989) *J. Construct. Steel Research*, **12**, 279–299.

(14) SMITH, I. F. C. and GURNEY, T. R. (1986) Changes in the fatigue life of plates with attachments due to geometrical effects, *Welding Res. Suppl.*, 244–250.

(15) NEWMAN, Jr., J. C., PHILLIPS E. P., SWAIN, M. H., and EVERETT, Jr., R. A. (1992) Fatigue mechanics: an assessment of a unified approach to life prediction, *Advances in fatigue lifetime predictive techniques, ASTM STP 1122*, ASTM Philadelphia, pp. 5–27.

G. Glinka* and S. Lambert*

Modelling of Fatigue Crack Growth in Flat Plate Weldments and Tubular Welded Joints

REFERENCE Glinka, G. and Lambert, S., **Modelling of fatigue crack growth in flat plate weldments and tubular welded joints.** *Fatigue Design*, ESIS 16 (Edited by J. Solin, G. Marquis, A. Siljander, and S. Sipilä) 1993, Mechanical Engineering Publications, London, pp. 73–93.

ABSTRACT Modelling of fatigue crack growth in 'T' butt plate weldments and tubular joints is discussed in the following paper. Some of the available stress intensity factor solutions based on the finite element method and the weight function approach are critically analysed. It is shown that simultaneous modelling of the fatigue growth of several multiple cracks provides a better simulation of crack shape development than do single crack models; single crack models may be unconservative.

It is also concluded that the flat plate stress intensity factor solutions can be used for cracks in tubular connections providing that the load shedding caused by the growing crack is accounted for.

Notation

a	Crack depth or minor semi-axis of semi-elliptical crack
Δa	Crack depth increment
a_i	Initial crack depth
a_f	Final crack depth
C	Paris' formula coefficient
c	Major semi-axis of semi-elliptical crack (crack length)
Δc	Crack length increment
da/dN	Crack growth rate
M_1, M_2, M_3	Weight function parameters
M_{1A}, M_{2B}, M_{3B}	Weight function parameters for the deepest point of semi-elliptical surface cracks
M_{1B}, M_{2B}, M_{3B}	Weight function parameters for the surface point of semi-elliptical surface crack
$m(x, a)$	Weight function
m_A	Weight function for the deepest point of semi-elliptical surface crack
m_B	Weight function for the surface point of semi-elliptical surface crack
m	Paris' formula exponent
N	Number of cycles
ΔN	Increment of number of cycles

* Department of Mechanical Engineering, University of Waterloo, Waterloo, Ontario, Canada N2L 3G1.

K	Stress intensity factor
K_A	Stress intensity factor at the deepest point of semi-elliptical surface crack
K_B	Stress intensity factor at the surface point of semi-elliptical surface crack
ΔK	Stress intensity range
t	Thickness
Y	Geometric stress intensity correction factor
α	Weld angle
ρ	Weld toe radius
σ_0	Nominal or reference stress
$\sigma(x)$	Stress distribution in the prospective crack plane
$\Delta\sigma_0$	Nominal stress range
σ_{HS}	Hot spot stress

Introduction

Usually, a large portion of the total fatigue life of a welded structure is spent during the propagation of a crack initiated at the weld toe **(1)(2)**. Therefore, fatigue and fracture analyses of notched and welded components are often carried out using fracture mechanics models. The prediction of the fatigue life for a welded joint requires accurate fatigue crack growth calculations in the highly-stressed region near the weld toe.

The modelling of fatigue crack growth in welded structures consists of several steps, the most important of which are:

– estimation of initial crack size and geometry;
– derivation of appropriate stress intensity factors;
– determination of stresses in the prospective crack plane;
– integration of fatigue crack growth rate formula;
– determination of the final or critical crack size and geometry.

Each of the five elements of fatigue crack growth analysis can be carried out in different ways depending on the geometry of a particular welded element, mode of loading, and available theoretical solutions.

The estimation of the initial crack size involves non-destructive crack detection and sizing **(3)(4)** or proof testing. The definition of the final or critical crack size depends on several engineering and safety aspects which are beyond the scope of this paper. In general, the choice of initial crack size has a much more significant influence on the life.

The analysis of fatigue crack growth from its initial dimensions to its final critical stage requires accurate calculations of instantaneous stress intensity factors and the integration of a suitable crack growth rate formula to determine crack growth increments. The calculations are complicated by the fact that edge or semi-elliptical surface cracks have to be analysed in a non-linear stress field. The methods of calculating stress intensity factors for cracks in weldments as well as numerical modelling of their growth due to cyclic loading is discussed below.

Stress intensity factors for edge and semi-elliptical cracks in 'T' butt weldments

Despite the fact that several stress intensity factor handbooks have been published (5)(6), it is still difficult to find adequate solutions for many welded configurations. This is mainly due to a wide variety of complex geometric and loading systems. It is known that the stress intensity factor for a crack in a welded joint depends on the overall geometry of the joint, the weld profile, and the type of loading. One of the frequently encountered geometrical configurations in welded structures are 'T' butt joints (Fig. 1) with fillet welds.

Edge crack in a 'T' butt welded joint

The stress intensity factors for an edge crack in the 'T' butt weldment, as shown in Fig. 1, have been determined by several authors (7)–(10) using the finite element method or approximate analytical solutions. The data are usually given in tabular or diagrammatical form in terms of the non-

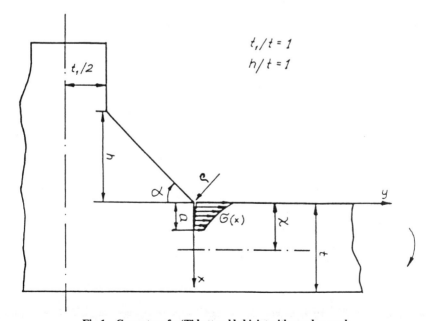

Fig 1 Geometry of a 'T' butt welded joint with an edge crack

dimensional geometric correction factor Y. The stress intensity factor is usually written in the form of equation (1) shown below.

$$K = \sigma_0 \sqrt{(\pi a)} Y \qquad (1)$$

Dimensionless stress intensity correction factors, Y, for an edge crack in a 'T' butt weldment under tension and bending load are given in Fig. 2. The data shown in Fig. 2 were obtained for one particular weld geometry characterized by the weld angle $\alpha = 45$ degrees and the weld toe radius $\rho/t = 1/25$. In numerical analyses, the weld toe radius ρ is not often analysed because the weld toe is modelled as a sharp corner. Such, an approach yields inaccurate stress intensity factors for small cracks, i.e., for $a < 3\rho$.

In order to calculate stress intensity factors for a variety of weld configurations, i.e., for different weld toe angles α and weld toe radii ρ, it is necessary

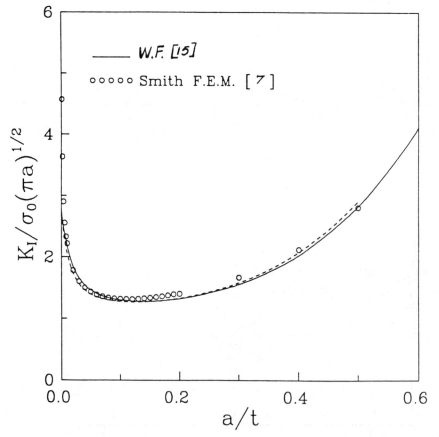

Fig 2(a)   Dimensionless stress intensity factor versus depth for an edge crack in a 'T' butt welded joint in tension

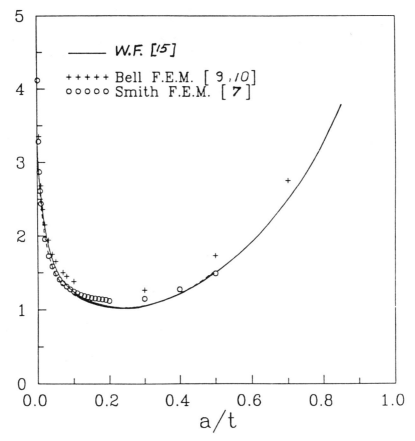

Fig 2(b) **Dimensionless stress intensity factor versus depth for an edge crack in a 'T' butt welded joint in bending**

to use the weight function approach **(11)(12)**. The unique feature of this method is that once the weight function for a particular cracked body is determined, the stress intensity factor for any loading system applied to the body can be calculated by a simple integration of the product of the stress field $\sigma(x)$ and the weight function $m(x, a)$.

$$K = \int_o^a \sigma(x)m(x_1a) \, \mathrm{d}x \qquad (2)$$

The stress field, $\sigma(x)$, has to be determined for an uncracked body using an analytical, numerical or experimental method of stress analysis. Typical through-the-thickness stress distributions in 'T' butt weldments are shown in Fig. 3. More data related to stress distributions in weldments with a variety of weld configurations can be found in references **(11)(13)(14)**.

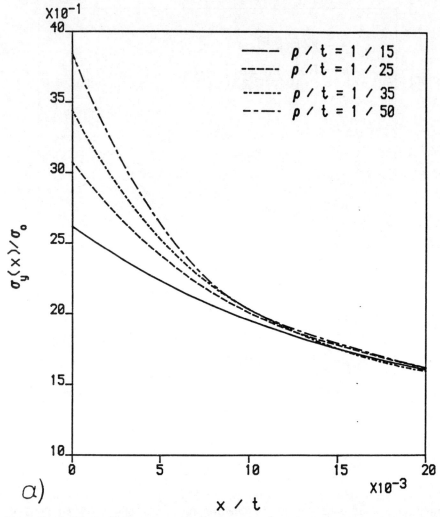

a)

Fig 3(a)   Through-the-thickness stress distribution in a 'T' butt weldment under bending load –
effect of the weld toe radius for $\alpha = 45$ degrees

The weight functions for any crack in mode I can be written in one general
form (15)

$$m(x, a) = \frac{2}{\sqrt{\{2\pi(a - x)\}}}$$

$$\times \left\{ 1 + M_1\left(1 - \frac{x}{a}\right)^{1/2} + M_2\left(1 - \frac{x}{a}\right) + M_3\left(1 - \frac{x}{a}\right)^{3/2} \right\} \qquad (3)$$

The parameters M_1, M_2, and M_3 depend on the geometry of a given cracked
body and can be found in references (15)–(17). It was also found (14) that in

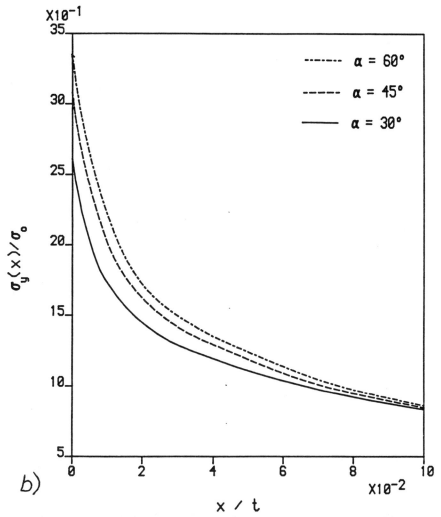

b)

Fig 3(b) Through-the-thickness stress distribution in a 'T' butt weldment under bending load –
effect of the weld angle for $\rho/t = 1/25$

the case of 'T' butt weldments the weight function derived for an edge crack in
a flat plate is applicable. The effect of the weldment geometry on the stress
intensity factor is accounted for by using equation (2) with the through-the-
thickness stress distribution $\sigma(x)$ in the prospective crack plane of the
uncracked weldment. Parameters M_1, M_2 and M_3 for an edge crack can be
found in the Appendix.

The effect of the weld angle α on the stress intensity factor obtained from
equation (2) by using the weight function (3) and the stress distribution from
Fig. 3(b) is shown in Fig. 4. The advantage of using the weight functions lies in

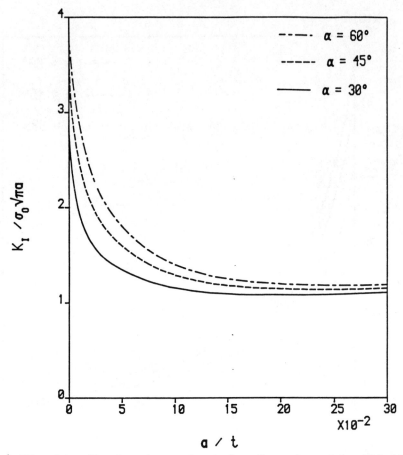

Fig 4   Effect of the weld angle on the stress intensity factor for an edge crack in a 'T' butt joint
($\rho/t = 1/25$) under bending load

the fact that the stress intensity factors can be easily calculated for any load configuration. The only input data required are the crack depth 'a' and the 'uncracked' stress distribution $\sigma(x)$.

Surface semi-elliptical cracks in a 'T' butt welded joint

The calculation of stress intensity factors for semi-elliptical surface cracks in weldments (Fig. 5) is difficult because a three-dimensional stress analysis is required for such cracks. Stress intensity factors for surface semi-elliptical cracks in plates are known only for a few simple load cases, i.e., Newman–Raju (18) solutions for pure bending and tension and Shiratori *et al.* (19) solutions for three exponential stress distributions. The most extensive numerical analysis of stress intensity factors for semi-elliptical cracks in weldments was carried out by Bell (9)(10). However, the data given in (9)(10) relate to only a

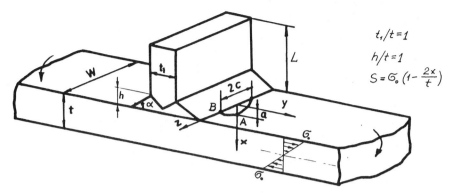

Fig 5   Semi-elliptical surface crack emanating from the weld toe of a 'T' butt welded joint

few geometrical weld configurations and two types of loading, i.e., pure bending and tension. Therefore, detailed analyses of several parameters such as plate thickness, weld thickness, weld angle, weld toe radius, and the loading system require a more efficient method than finite element or analytical approaches. Once again it was found that the weight function approach was sufficiently accurate and efficient.

It is also known that the growth of a surface semi-elliptical crack can be satisfactorily characterized by the extension of its two semi-axes. Therefore, only two weight functions need be derived **(17)**, i.e., one for the deepest point A (Fig. 5) and the second for the surface point B.

$$m_A = \frac{2}{\sqrt{\{2\pi(a-x)\}}}$$

$$\times \left\{ 1 + M_{1A}\left(1 - \frac{x}{a}\right)^{1/2} + M_{2A}\left(1 - \frac{x}{a}\right) + M_{3A}\left(1 - \frac{x}{a}\right)^{3/2} \right\} \qquad (4)$$

$$m_B = \frac{2}{\sqrt{(\pi x)}} \left\{ 1 + M_{1B}\left(\frac{x}{a}\right)^{1/2} + M_{2B}\left(\frac{x}{a}\right) + M_{3B}\left(\frac{x}{a}\right)^{3/2} \right\} \qquad (5)$$

The geometric parameters M_{1A}, M_{2A}, M_{3A}, M_{1B}, M_{2B}, and M_{3B} are given in the Appendix.

The weight functions (4) and (5) were originally derived for semi-elliptical cracks in plates but it was found **(12)(14)** that they could also be used for cracks in weldments. A typical set of data obtained by integration of weight function (4) and the stress distribution from Fig. 3(b) is given in the form of the non-dimensional geometric factor Y in Fig. 6. The stress intensity factors depend on the stress field $\sigma(x)$ and on two geometrical parameters, i.e., the crack shape and depth, characterized by ratios a/c and a/t, respectively.

The modelling of fatigue crack growth of semi-elliptical cracks requires the calculation of stress intensity factors for a variety of a/c and a/t values which change after each crack growth increment. Therefore, the geometric correction

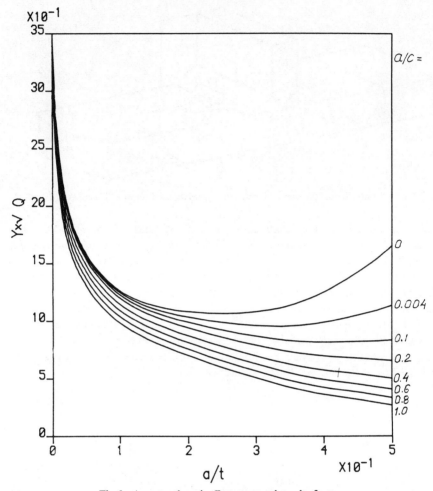

Fig 6   Aspect ration a/c effect on stress intensity factor

factors Y calculated earlier for several fixed a/c values (Fig. 6) are not sufficient because modelling of fatigue crack growth requires the calculation of stress intensity factors for all instantaneous values of a/c and a/t which may occur during crack growth. For this reason the weight function approach seems to be very useful since it enables the stress intensity factors for any values of a/c and a/t to be readily calculated.

Modelling of fatigue crack growth in 'T' butt weldments

Single edge cracks

The fatigue crack growth modelling is usually based on the Paris formula in the form

$$\frac{\mathrm{d}a}{\mathrm{d}N} = C(\Delta K)^m \tag{6}$$

In most cases relevant for welded joints, the effects of the mean stress is omitted. It is argued that due to a relatively high stress concentration and the presence of residual stresses the mean stress effect is minimized. Therefore, it is sufficient to relate the crack growth rate to the stress intensity range ΔK using equation (6).

Thus, in the case of an edge crack, the integration of equation (6) gives the number of cycles necessary to extend the crack from its initial size a_i to its final dimension a_f. The numerical integration technique involving equation (7) can be easily optimized because the stress intensity factor range ΔK represented by the Y parameter (Fig. 2) is known before the integration begins.

$$N = \int_{a_i}^{a_f} \frac{\mathrm{d}a}{C(\Delta K)^m} \tag{7}$$

However, the edge cracks usually become relevant only during the final stage of fatigue life of welded joints. Most of the fatigue life of welded joints is spent on propagation of semi-elliptical surface cracks.

Single semi-elliptical surface cracks

In order to model the growth of semi-elliptical cracks, it is necessary to analyse at least two factors: the shape of the crack and its depth characterized by parameters a/c and a/t, respectively. These are two methods in common use.

The first method is based on using the stress intensity factor corresponding to the deepest point A (Fig. 5) and calculating crack depth increments from equation (7). The instantaneous value of the parameter a/c characterizing the crack shape is calculated from the experimentally-determined forcing function, one example of which is [20]

$$a/c = e^{-ka} \tag{8}$$

The empirical constant k was found [20] to be dependent on the applied stress range $\Delta\sigma_0$, the material and the plate thickness. In the case of 'T' butt joints made of BS 4360 : 50D steel the k value can be estimated from the following formula [20]

$$k = 2.915 \times 10^{-6}(\Delta\sigma_0)^2\sqrt{t} \tag{9}$$

where a and t are given in mm and the stress range $\Delta\sigma_0$ in MPa.

This approach, as shown in Fig. 7, gives reasonable agreement between the calculated and experimentally-measured crack growth. The disadvantage of this method lies in the fact that the constant k in equation (8) has to be determined experimentally and, therefore, its applicability is limited to one particular material and geometry.

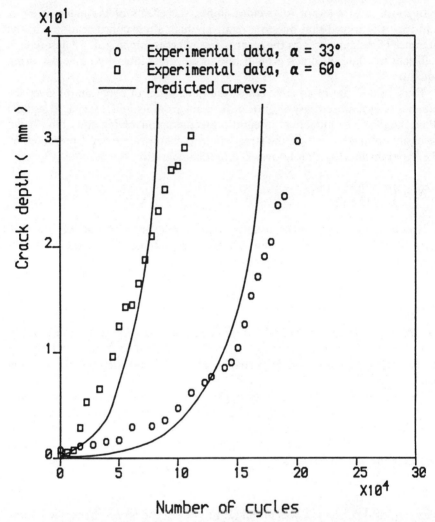

Fig 7 Predicted and calculated fatigue crack growth based on the deepest point stress intensity factor and the experimental shape forcing function, bending load $a_i = 0.2$ mm, $\rho = 1.4$ mm, $t = 70$ mm, $\Delta\sigma_0 = 175$ MPa

The second method of modelling the growth of semi-elliptical cracks in weldments is based on calculating two ranges of stress intensity factors ΔK_A and ΔK_B corresponding to the deepest point A (Fig. 5) and the surface point B on the crack front. Then, the crack growth increments are calculated in both directions using the Paris formula (equation (6)).

$$\Delta a = C(\Delta K_A)^m \Delta N \tag{10}$$

$$\Delta c = C(\Delta K_B)^m \Delta N \tag{11}$$

The calculation of stress intensity factors in equations (10) and (11) could be based on the integration of weight functions (4) and (5).

It is shown in references **(20)(21)** that the analysis based on equations (10) and (11) often yields non-conservative fatigue life predictions. This is due to the fact that the initial stages of the fatigue life of a welded component involves initiation, growth, and coalescence of several small elliptical cracks, distributed randomly along the weld toe. The analysis based on the modelling of single crack growth, without the use of a forcing function, neglects the effect of the existence of multiple cracks before the leading single crack is developed. Therefore, crack shape development is not modelled with sufficient accuracy.

Multiple semi-elliptical surface cracks

Recently, several attempts have been made **(22)** to model the simultaneous growth and coalescence of multiple semi-elliptical cracks appearing along the weld toe. This requires simultaneous calculations of fatigue crack growth increments for several semi-elliptical cracks of different size and geometry by using the set of two equations (10) and (11) for each crack. Such an approach involves many calculations because the stress intensity factors ΔK_A and ΔK_B have to be calculated for each crack separately. Once again, the weight function method proved to be the most suitable. It was found **(22)** that the interaction between adjacent cracks could be neglected in the calculation of the stress intensity factors for each crack. Coalescence was assumed to occur instantaneously once adjacent cracks touched at the surface, forming a single semi-elliptic crack. Comparison of the calculated and experimentally measured fatigue crack evolution in a 'T' butt weldment under bending load **(22)** is shown in Fig. 8.

Modelling of fatigue crack growth in tubular welded joints

In simple tubular joints, fatigue cracks usually initiate as surface cracks from the weld toe and propagate (Fig. 9) into the tubular wall. The main problem in the modelling of fatigue cracks in tubular joints compared with 'T' butt weldments is the calculation of the stress intensity factor. The calibration of stress intensity factor solutions requires three dimensional linear–elastic stress analyses of cracked tubular joints. Analytical solutions to such problems are not available due to the mathematical complexity of the geometry and loading. Since three-dimensional finite element analyses of cracked tubular joints are very time consuming, expensive, and cumbersome, experimentally calibrated stress intensity factors **(23)** or approximate analytical solutions **(24)(25)** have been used. The available approximate solutions are normally derived for surface cracks in plate weldments. However, it is known that stress intensity factors for semi-elliptical cracks in thin wall cylinders are not significantly different **(26)** from flat plate solutions. Therefore, the stress intensity

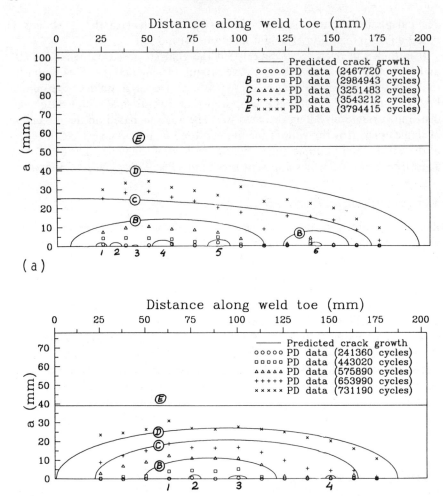

Fig 8 **Comparison of experimentally measured and calculated fatigue crack evolution using the multiple crack growth modelling: 'T' butt joint under bending** (22)

factor solutions derived for flat plate weldments, such as 'T' butt joints, appear to be sufficiently accurate for application to tubular joints.

It is also important to note that calculations of stress intensity factors for weldments made of flat plate elements are based on the reference or nominal stresses σ_0, which are usually assumed to be the average stress in pure tension or the bending stress, calculated from simple beam theory. Calculation of such stresses in tubular joints is rather difficult because the stresses near the tube intersections are not simply related to the applied load as it is in the case of

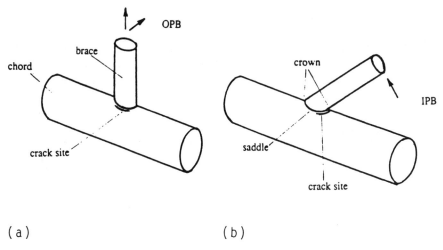

(a)   (b)

**Fig 9   Tubular joints and modes of loading. (a) 'T' joint subjected to axial or out-of-plane bending.
(b) 'Y' joint subjected to in-plane bending**

plate 'T' butt weldment. Therefore, the basic reference stress used in the design
of tubular welded joint is the 'hot spot stress', σ_{HS}, as defined in reference **(27)**
and illustrated in Fig. 10. It is worth noting that the hot spot stress, σ_{HS}, is not
the true stress at the weld toe as it neglects the peak stress σ_{peak} resulting from
the stress concentration due to the local weld geometry. Thus, the definition of

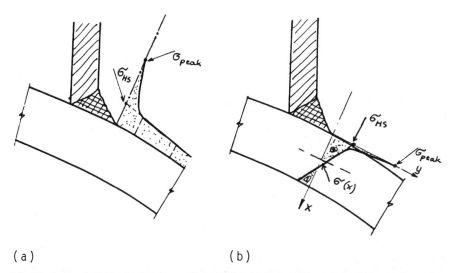

(a)   (b)

**Fig 10   Stress variation close to the weld toe of tubular intersection and the 'hot spot stress' σ_{HS}
definition (a) on exterior tubular surface and (b) through-the-wall thickness**

the stress concentration factor for tubular welded joints and the stress intensity factor employ the hot spot stress, σ_{HS}, as the reference or nominal stress.

All of the stress intensity factor solutions and the weight functions presented above have been derived for the case of cracks in bodies which are basically statically determinate structures in which the reference or nominal stress σ_0 does not depend on the local stiffness of the cracked section. For this type of structure it is possible to use the weight function and the superposition technique for the integration of equation (2). In the case of tubular joints the reference hot-spot stress, σ_{HS}, depends on the local stiffness which changes with the growing crack. Therefore, it was found (28)(29) that the stress intensity factors obtained for plate weldments (Fig. 6), or those obtained from integration of the weight function with constant hot-spot stress, σ_{HS}, in equation (2) resulted in a significant overestimation of stress intensity factors. A comparison of experimentally-determined and calculated stress intensity factor for tubular 'T' joint is shown in Fig. 11. The theoretical data shown in Fig. 11 were obtained from equation (2) and the weight function (4). The stress distribution normalized with respect to the hot-spot stress, σ_{HS}, was determined using the finite element method. It is apparent that the stress intensity factors calculated on the basis of constant σ_{HS} stress were significantly higher than the experimental ones. On the other hand, it is apparent that in the case of plate 'T' butt weldment the stress intensity factors determined from the same weight function (2) agreed very well with the accurate numerical analysis, which was used instead of experiments. Therefore, it was concluded that in the case of tubular joints more information is needed for the calculation of stress intensity factors than the weight function and the initial stress distribution in the uncracked section.

The major differences between a crack in a plate welded joint and similar cracks in a tubular joint are due to different boundary conditions. Equations (2) and (4) are valid for statically determinant configurations where the same moments and normal loads were transferred through the cracked section. In the case of tubular joints, the crown/saddle bending moments and membrane stresses driving the crack depend on the cracked section stiffness and as a consequence they are dependent on the crack size. Therefore, it was concluded (28)(30) that using the constant stress input in equation (2) representing the initial stress state in an uncracked section is an overestimation of the effective stress level driving the crack growth in tubular joints. Consequently, it was argued (28)(29) that due to decreasing stiffness, the cracked section in a tubular joint transferred lower loads than the uncracked section. In the case of a plate joint such as the 'T' butt weldment (Fig. 5) the load transferred through the cracked section was always constant and independent of the cracked section stiffness and as a result independent of the crack size. The decrease in the load transferred through the cracked section in a tubular joint was called the 'load shedding' effect.

To study the effects of the boundary conditions influencing the load shedding a numerical analysis of a series of flat plates and rings containing edge

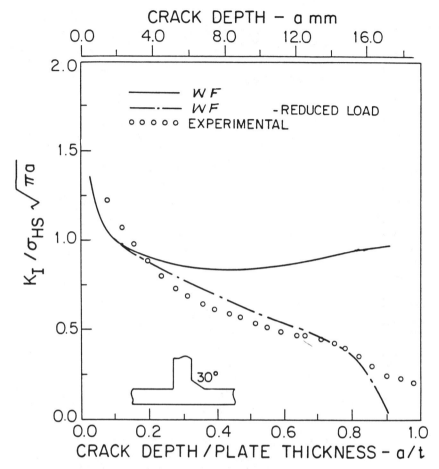

Fig 11 Experimental and theoretical calibration of the geometrical stress intensity factor: 'T' joint under cyclic tension load (29)

cracks was presented in reference **(30)**. Recently, experimental data **(29)** has been obtained to determine the characteristics of the load shedding. The data shown in Fig. 11 clearly indicate that the use of the decreasing reference stress, σ_{HS}, characterizing the stress distribution used in equation (2) improved the agreement between the calculated and experimental stress intensity factors. Similar improvement was achieved by using a simple linear model to decrease the hot spot stress, σ_{HS}, **(28)** with increasing crack depth a. The effect of the load shedding on the predicted fatigue crack propagation life is shown in Fig. 12. The fatigue crack growth lives were predicted on the basis of an experimentally-determined forcing function (equation (8)) and the stress intensity calculated for the deepest crack point A. It is apparent that inclusion of the load shedding model improved the estimation of the fatigue life.

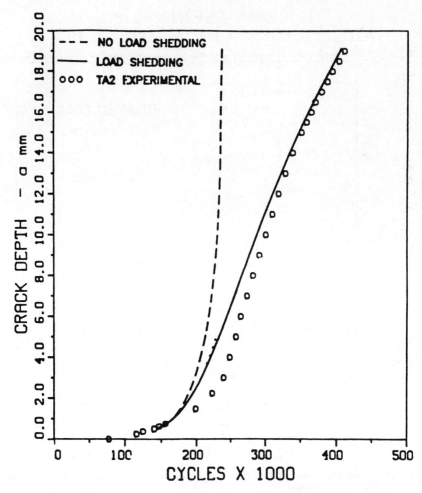

Fig 12 **The effect of load shedding on fatigue crack growth in a tubular 'T' joint under in tension** (13)

Conclusions

It has been shown that the stress intensity factors for edge and surface semi-elliptical cracks in weldments can be calculated using weight functions. These functions allow the stress intensity factor to be analysed with respect to several geometrical and loading parameters such as weld toe, radius, weld angle, plate thickness, and combination of several loading inputs.

It was found that the fatigue life estimations of flat plate weldments such as 'T' butt joints were non-conservative while based on a single semi-elliptical crack model with stress intensity factors calculated for the deepest and surface crack points. Better estimations of fatigue lives were achieved based on ·the single crack model using the stress intensity factor at the deepest point and the experimental crack shape forcing function. An alternative approach, which allows for the presence of multiple semi-elliptic cracks also shows some promise.

It has been found that in the case of tubular joints it was necessary to supplement the methodology developed for plate weldments by an additional effect called 'load shedding'. It is believed that the load shedding effect arises due to varying boundary conditions caused by the growing crack.

Appendix

Weight function parameters for a single edge crack in a finite thickness plate (equation (3))

$$M_1 = -0.1436 - 1.1731(a/t) + 64.1625(a/t)^2 - 685.562(a/t)^3$$
$$+ 3281.01(a/t)^4 - 8471.29(a/t)^5 + 12\,213.1(a/t)^6$$
$$- 9267.91(a/t)^7 + 2898.2(a/t)^8;$$

$$M_2 = 0.96\,049 - 1.1960(a/t) - 35.2818(a/t)^2 + 794.362(a/t)^3$$
$$- 4359.98(a/t)^4 + 11\,933.1(a/t)^5 - 17\,575.7(a/t)^6$$
$$+ 13\,399.4(a/t)^7 - 4161.42(a/t)^8;$$

$$M_3 = -0.0129 - 35.5956(a/t) + 956.47(a/t)^2 - 9196.73(a/t)^3$$
$$+ 44\,136.2(a/t)^4 - 116\,851.0(a/t)^5 + 173\,971.0(a/t)^6$$
$$- 1\,365\,24.0(a/t)^7 + 44\,068.6(a/t)^8;$$

Weight function parameters for a surface (Fig. 5) semi-elliptical crack in a finite thickness plate (valid for $0.2 \leqslant a/c \leqslant 1$ and $0 \leqslant a/t \leqslant 0.8$)

For the deepest point of (equation (4))

$$M_{1A} = \frac{\pi}{\sqrt{(2Q)}}(4Y_0 - 6Y_1) - \tfrac{24}{5}; \quad M_{2A} = 3;$$

$$M_{3A} = 2\left(\frac{\pi}{\sqrt{(2Q)}}Y_0 - M_{1A} - 4\right);$$

$$Y_0 = B_0 + B_1\left(\frac{a}{t}\right)^2 + B_2\left(\frac{a}{t}\right)^4,$$

$$B_0 = 1.10\,190 - 0.019\,863(a/c) - 0.043\,588(a/c)^2,$$
$$B_1 = 4.32\,489 - 14.9372(a/c) + 19.4389(a/c)^2 - 8.52\,318(a/c)^3,$$
$$B_2 = -3.03\,329 + 9.96\,083(a/c) - 12.582(a/c)^2 + 5.3462(a/c)^3,$$

$$Y_1 = A_0 + A_1(a/t)^2 + A_2(a/t)^4,$$

$$A_0 = 0.456\,128 - 0.114\,206(a/c) - 0.046\,523(a/c)^2,$$

$$A_1 = 3.022 - 10.8679(a/c) + 14.94(a/c)^2 - 6.8537(a/c)^3,$$

$$A_2 = -2.28\,655 + 7.88\,771(a/c) - 11.0675(a/c)^2 + 5.16\,354(a/c)^3$$

$$Q = 1. + 1.464(a/c)^{1.65} \; for \; 0 \le a/c \le 1$$

For the surface point B (equation (5))

$$M_{1B} = \frac{\pi}{\sqrt{(4Q)}} (30F_1 - 18F_0) - 8; \; M_{2B} = \frac{\pi}{\sqrt{(4Q)}} (60F_0 - 90F_1) + 15;$$

$$M_{3B} = -(1 + M_{1B} + M_{2B})$$

where

$$F_0 = \alpha(a/c)^\beta$$

$$\alpha = 1.14\,326 + 0.0\,175\,996(a/t) + 0.501\,001(a/t)^2$$

$$\beta = 0.458\,320 - 0.102\,985(a/t) - 0.398\,175(a/t)^2$$

$$F_1 = \gamma(a/c)^\delta$$

$$\gamma = 0.976\,770 - 0.131\,975(a/t) + 0.484\,875(a/t)^2$$

$$\delta = 0.448\,863 - 0.173\,295(a/t) - 0.267\,775(a/t)^2$$

References

(1) JAKUBCZAK, H. and GLINKA, G. (1986) Fatigue analysis of manufacturing defects in weldments, *Int. J. Fatigue*, **8**, 51–57.
(2) YEE, R., BURNS, D. J., MOHAUPT, U. H., BELL, R., and VOSIKOVSKY, O. (1990) Thickness effect and fatigue crack development in welded T-joints, *J. Offshore Mech. Arctic Engng*, **12**, 341–351.
(3) NEWTON, K. (1987) The transparency of fatigue cracks to NDT methods used for the inspection of offshore structures, Offshore Europe 87 – Conference Proceedings, p 1–17.
(4) SMITH, I. F. C. (1983) Detection and measurement of fatigue cracks in welded joints, *Sixth ASM International Conference on NDE in the Nuclear Industry*, Vol. 28 pp. 789–795.
(5) TADA, M., PARIS, P., and IRWIN, G. (1973) *The stress analysis of cracks handbook*, Del Research Corporation, USA.
(6) ROOKE, D. P. and CARTWRIGHT, D. J. (1976) *Compendium of stress intensity factors*, Hillington Press, Uxbridge.
(7) SMITH, I. F. (1984) The effect of geometry changes upon the predicted fatigue strength of welded joints, *Proceedings of the third International Conference on Numerical Methods in Fracture Mechanics*, (Edited by A. R. Luxmore and D. R. J. Owen), Pineridge Press, Swansea, UK, pp. 561–574.
(8) SKORUPA, M., BRAAM, H. and PRIJ, J. (1987) Applicability of approximate K_1 solutions towards cracks at weld toes, *Engng Fracture Mech.*, **26**, 669–691.
(9) BELL, R. and KIRKHOPE, J. (1984) Determination of stress intensity factors for weld toe defects, Final Report, DSS Contract OSU82–00233, Carleton University, Ottawa, Canada.
(10) BELL, R. (1985) Determination of stress intensity factors for weld tow-defects-phase II, Final Report, DSS Contract OST84–00125, Carleton University, Ottawa, Canada.

(11) NIU, X and GLINKA, G. (1987) The weld profile effect on stress intensity factors in weldments, *Int. J. Fracture*, **35**, 3–20.

(12) NIU, X. and GLINKA, G. (1989) Stress intensity factors for semi-elliptical suface cracks in welded joints, *Int. J. Fracture*, **40**, 255–270.

(13) FORBES, J. W. (1991) *Fatigue in stiffened T-tubular joints for offshore structures*, PhD Thesis, University of Waterloo, Canada.

(14) FORBES, J. W., DESJARDINS, J. L., GLINKA, G. and BURNS, D. J. (1991) Stress intensity factors for semi-elliptical surface cracks in weldments, *Proceedings of the 10th International Conference on Offshore Mechanics and Arctic Engineering*, (Edited by M. Salama and R. Chong), ASME, USA, pp. 529–536.

(15) GLINKA, G. and SHEN, G. (1991) Universal features of weight functions for cracks in mode I, *Engng Fracture Mech.*, **4**, 1135–1146.

(16) SHEN, G. and GLINKA, G. (1991) Determination of weight function from reference stress intensity factors, *Theoretic Appl. Fracture Mech.*, **15**, 237–245.

(17) SHEN, G. and GLINKA, G. (1991) Weight functions for a surface semi-elliptical crack in a finite thickness plate, *Theoret. Appl. Fracture Mech.*, **15**, 247–255.

(18) NEWMAN, J. C. and RAJU, I. S. (1986) Stress intensity factors equations for cracks in three dimensional finite bodies subjected to tension and bending loads, *Computational Methods in the Mechanics of Fracture*, (Edited by S. N. Atluri), North-Holland, Holland, pp. 311–334.

(19) SHIRATORI, M., NIYOSHI, T., and TANIKAWA, K. (1987) Analysis of stress intensity factors for surface cracks subjected to arbitrarily distributed surface stress, *Stress intensity factors handbook*, (Edited by Y. Murakami, *et al.*) Vol. 2, Pergamon Press, Oxford, pp. 698–705.

(20) VOSIKOVSKY, O., BELL, R., BURNS, D. J., and MOHAUPT, U. H. (1985) Fracture mechanics assessment of fatigue life of welded plate T-joints, including thickness effects, *Proceedings of 4th International Conference on Behaviour of Offshore Structures*, Elsevier, pp. 453–464.

(21) BELL, R., VOSIKOVSKY, O., BURNS, D. J., and MOHAUPT, V. H. (1987) A fracture mechanics model for life prediction of welded plate joints, *Proceedings of the 3rd ECSC Offshore Conference on Steel in Marine Structures*, (Edited by C. Noordhoek and J. de Back) Elsevier, pp. 901–910.

(22) LAMBERT, S. B. and BURNS, D. J. (1993) A multiple crack model for fatigue in welded joints, *Int. J. Fatigue*, **15**.

(23) DOVER, W. D. and DHARMAVASAN, S. (1982) Fatigue fracture mechanics analysis of T and Y joints, Proceedings of the 14th Offshore Technology Conference, Paper No. OTC 4404.

(24) DOVER, W. D. and CONNOLLY, M. P. (1986) Fatigue fracture mechanics assessment of tubular welded Y and K joints, *International Conference on Fatigue and Crack Growth in Offshore Structures*, (Edited by W. D. Dover, G. Glinka, and A. Reynolds), I MechE, London, pp. 117–135.

(25) THORPE, T. N. (1987) A simple model for fatigue prediction and in-service defect assessment of tubular joints, Proceedings of the 3rd International Symposium on Integrity of Offshore Structures, pp. 285–305.

(26) FORMAN, R. G. and SHIVAKUMAR, V. (1984) Growth behaviour of surface cracks in the circumferential plane of solids and cylinders, *The 17th National Symposium on Fracture Mechanics, ASTM STP 905*, ASTM Philadelphia, pp. 54–75.

(27) IRVINE, N. M. (1981) Review of stress analysis techniques in UKOSRP, Proceedings of the Conference on Fatigue in Offshore Structural Steels, pp. 47–57.

(28) AAGHAAKOUCHAK, A., GLINKA, G., and DHARMAVASAN, S. (1989) A load shedding model for fracture mechanics analysis of fatigue cracks in tubular joints, *Proceedings of the 8th International Conference on Offshore Mechanics and Arctic Engineering*, (Edited by M. Salama *et al.*), ASME, pp. 159–165.

(29) FORBES, J., GLINKA, G., and BURNS, D. J. (1992) Fracture mechanics analysis of fatigue cracks and load shedding in tubular welded joints, *Proceedings of the 11th International Conference on Offshore Mechanics and Arctic Engineering*, (Edited by M. Salama *et al.*), ASME.

(30) AAGHAAKOUCHAK, A., DHARMAVASAN, S., and GLINKA, G. (1990) Stress intensity factors for cracks in structures under different boundary conditions, *Engng Fracture Mech.*, **37**, 1125–1137.

H. Kawono* and K. Inoue*

A Local Approach for the Fatigue Strength Evaluation of Ship Structures

REFERENCE Kawono, H. and Inoue, K., **A local approach for fatigue strength evaluation of ship structures,** *Fatigue Design.* ESIS 16, (Edited by J. Solin, G. Marquis, A. Siljander, and S. Sipilä) 1993, Mechanical Engineering Publications, London, pp. 95–108.

ABSTRACT The direct strength calculation method has been increasingly used in ship structural design. It is not, however, clear how FEM stress distribution ought to be used in the assessment of fatigue strength of welded components. By concentrating on the round fillet weld, the authors define a critical crack configuration to be assumed in the design stage. Then, by using an assumed parameter $C(r)$, the authors analyse the fatigue test data, and consequently find that the $C(r)$ value can commonly apply in these instances within approximately 5 mm of the weld toe. It is, therefore, concluded that the local stress at 5 mm from the weld toe would be appropriate for practical use as a reference stress to estimate the fatigue strength of round fillet weld.

Introduction

In ship structural design, the direct strength calculation method, based on FEM stress calculations, is commonly used to determine the dimensions of the members.

In the evaluation of fatigue strength, the stress distribution around the spots in which fatigue cracks may occur (hot spots) is also given by FEM. However, the following problems continually arise.

(a) Detailed stress distribution can be shown by FEM; however, it is difficult to utilize the results in fatigue estimation. This problem relates to the stress axis in S–N diagrams. There are, however, also indefinite points relating to the fatigue life axis.

(b) How are life of a text specimen and the crack condition of an actual ship correlated?

The above problem seems to be fundamental in attaining stress calculation and fatigue estimation. This is consistent with the direct strength calculation method. The research results by SRAJ (5) and IIW documents (1)(3)(4) etc. are available, and the joint research by SRAJ (6) is in progress. This paper focuses on the round fillet weld which is very common in ship structures. The obtained result on the fatigue crack behaviour and the analysis of the experimental data are described (2).

* Nagasaki Research and Development Center, Mitsubishi Heavy Industries Limited, Nagasaki, Japan.

Characteristics of fatigue crack growth behaviour in a round fillet weld

Fatigue crack initiation on the actual ship mostly occurs at the toes of a round fillet weld. It is, therefore important to establish the estimation procedure of the fatigue strength in these places. Figure 1 shows an example of a crack growth diagram in an elementary joint with a longitudinal rib under constant load amplitude. This figure illustrates that the crack can be recognized considerably early compared with the failure life $N_f(N/N_f = 0.05\text{-}0.1$ in Fig. 1). It can also be seen that, measured by the semi-logarithmic scale, the crack growth line suddenly rises, like the tip of shoe at a certain life level, flattens in the slope, and then reaches its final breaking point.

This corresponds to the fact that small cracks are simultaneously generated along the toe line of reinforcement of a weld connected by growth. The number of cycles increases just before the crack deviates from the toe line. The crack configuration in which the crack tip deviates toward the base metal corresponds to l_{cr} as illustrated in Fig. 2.

In view of the above crack growth phenomenon and the detectability of the cracks under the actual ship condition, the following crack growth condition should be assumed as the design criteria. In other words, the life N_D at the time when the crack tip is just separating from the toe is used as the critical

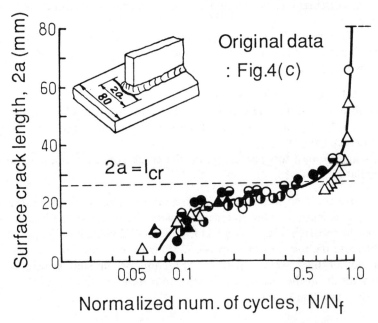

Fig 1   Example of crack growth diagram at toe of round fillet weld

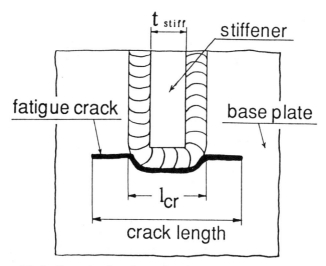

Fig 2   Conceptional crack growth path at toe of round fillet weld

crack configuration in designing.

$$N_D = N(l = l_{cr}) \tag{1}$$

$$l_{cr} = t_{stiff} + 2 \times \text{(leg length)} \cong 2 \times t_{stiff} \tag{2}$$

As above, when the life N_D is determined, it is found that N_D includes the life n_i until crack is generated and the life N_P for growth and expansion of the crack. Generally the higher the stress concentration, the larger the governing power of the propagation life N_P. The propagation life N_P becomes important where stress is highly concentrated, such as at a round fillet weld toe.

Consideration regarding fatigue estimating stress

Although local stress can be obtained in considerable detail by FEM, the calculation effort increases significantly to take the detailed shape of weld reinforcement into consideration. Moreover, it is unavoidable that the toe radius of a weld reinforcement, bead shape etc. can be very varied. The FEM modelling, such as toe radius, is not impossible in principle; however the practical significance of the solution obtained and the contribution to fatigue estimation is questionable. In order not to be affected by the detailed shape of the weld, the stress information at a certain distance apart from the toe of weld reinforcement must be used, while in hull members the as-welded condition is standard practice.

Therefore re-analysis on the above fatigue test data was carried out in accordance with the following basic considerations; the definition of fatigue estimating stress has been considered.

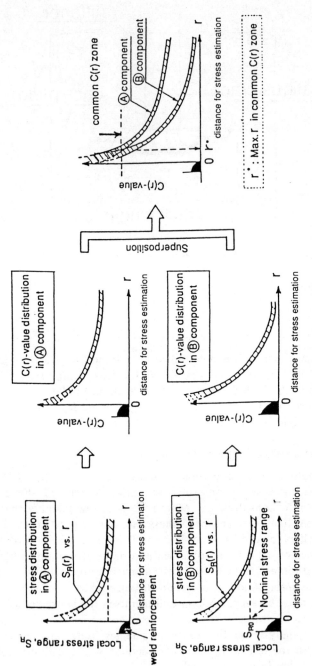

Fig 3 Conceptional flow for estimating stress definition

In order to study stress relating to high cycle fatigue, the parameter $C(r)$ (r : distance from toe of weld reinforcement) defined by the following assumptions is introduced.

(a) The fatigue strength diagram is assumed to be expressed by equation (3). The local stress range S_R is a function of position around the hot spot. C value in the right side of the equation is assumed to be a function of position as well as S_R.

$$S_R \times N_D^\alpha = C(r), \quad (\alpha : \text{experimental constant}) \tag{3}$$

(b) Although the shapes near the round fillet weld are different (e.g., whether snipping end or not), the exponent α in the above equation is assumed to be the common value if the welding procedure of the round fillet weld is same.

Whether the concept of the fatigue estimating stress in Fig. 3 is acceptable or not depends on whether $C(r) = $ const. is commonly satisfied among many kinds of fatigue tests within a certain distance from the toe of weld reinforcement. The acceptable range of this approach, however, is determined by the range of the test data.

The above analysis has been applied to the results of the fatigue test series with a round fillet weld (see Figs 4–6).

It is worth noting that the fatigue crack of the end bracket model started from the toe of the bracket side out of two weld toes (base plate side/bracket side), and that stress decrease extremely as being away from toe of weld reinforcement.

The fatigue life N_D is determined using the results in the preceding section. The local stress range distribution around hot spot is plotted mainly by using the superficially measured strain multiplied by modulus of elasticity and estimated by extrapolation and interpolation. The experimental constant in

No.	material	stress range, S_{R0} (kgf/mm^2[MPa])	failure life, Nf (cycles)
H-1		19.6 [192.]	$1.47*10^5$
H-2		29.4 [288.]	$4.23*10^4$
H-3	KA36	14.7 [144.]	$3.10*10^5$
H-4		24.5 [240.]	$8.36*10^4$
H-5		12.2 [120.]	$8.85*10^5$
H-6		9.8 [96.]	$1.52*10^6$

Fig 4(a)   Fatigue data by elementary joint specimen with longitudinal rib – summary of test data

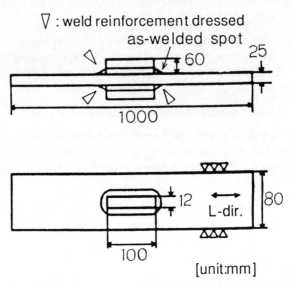

Fig 4(b)   Fatigue data by elementary joint specimen with longitudinal rib – shape of specimen

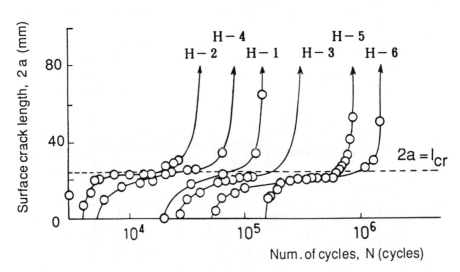

Fig 4(c)   Fatigue data by elementary joint specimen with longitudinal rib – crack growth curve

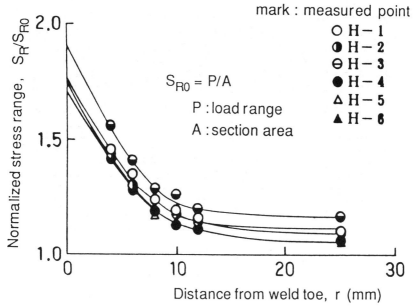

Fig 4(d) Fatigue data by elementary joint specimen with longitudinal rib – measured local stress range distribution

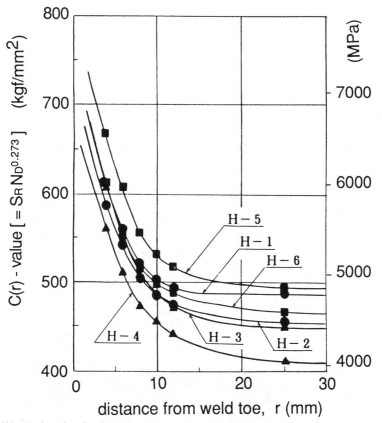

Fig 4(e) Fatigue data by elementary joint specimen with longitudinal rib – C(r) versus r diagram

specimen No.	load, P (ton) MIN.	load, P (ton) MAX.	stress range, S_{R0} (kgf/mm^2[MPa])	failure life, Nf (cycles)
A-1	2.5	25.0	11.3 [110.]	$7.44*10^5$
A-2	3.5	35.0	16.0 [157.]	$2.11*10^5$
A-3	1.8	18.0	8.2 [80.]	$1.94*10^6$
A-4	1.5	14.5	6.8 [66.]	$3.36*10^6$
A-5	4.3	43.0	20.0 [196.]	$9.27*10^4$
A-6	2.0	20.0	9.1 [89.]	$1.48*10^6$

Fig 5(a)   Fatigue data by small bracket model – summary of test data

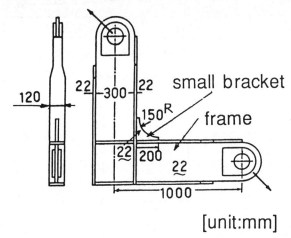

[unit:mm]

Fig 5(b)   Fatigue data by small bracket model – shape of test model

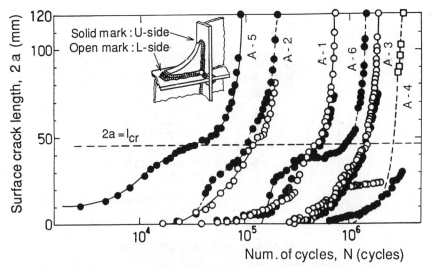

Fig 5(c)   Fatigue data by small bracket model – crack growth curve

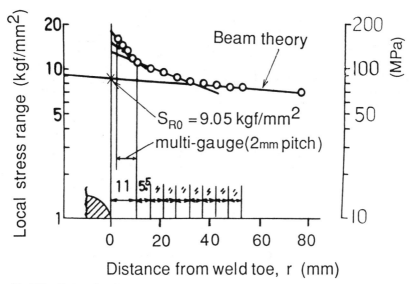

Fig 5(d) Fatigue data by small bracket model – measured local stress range distribution

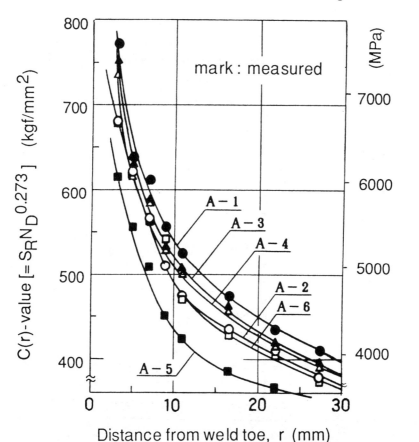

Fig 5(e) Fatigue data by small bracket model – $C(r)$ versus r diagram

specimen No.	load, P (ton)		nominal stress range S_{R0} (kgf/mm2[MPa])	failure life Nf (cycles)	note*
	MIN.	MAX.			
No. 1	2.0	20.0	11.9 [117.]	$1.386*10^5$	B
No. 2	0.7	7.2	12.6 [123.]	$0.736*10^5$	B
No. 3	2.6	26.0	12.9 [126.]	$2.230*10^5$	B (A)
No. 4	2.3	23.0	12.0 [118.]	$1.456*10^5$	B

*   A : crack from the toe of base plate side

      B : crack from the toe of bracket side

Fig 6(a)   Fatigue data by end bracket model – summary of test data

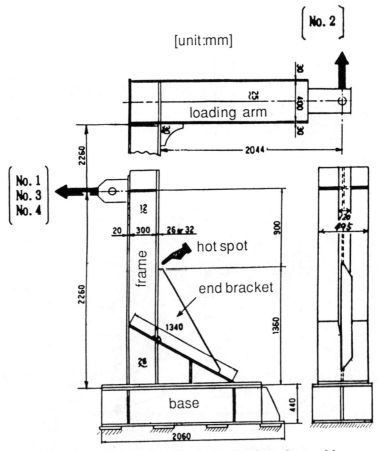

Fig 6(b)   Fatigue data by end bracket model – shape of test model

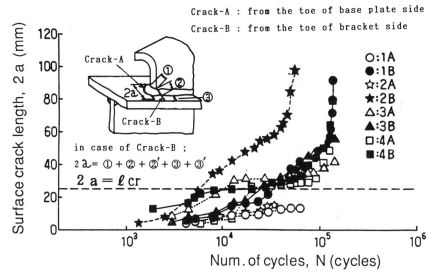

Fig 6(c)   Fatigue data by end bracket model – crack growth curve

equation (3) was set by $\alpha = 0.273$, which was obtained from the test result of the SN diagram (nominal stress range versus failure life) shown in Fig. 4(a). Making sure the applicability of the conceptual flow in Fig. 3, the results in Fig. 4(e)–Fig. 6(e) were re-written and shown in Fig. 7.

As clearly shown in Fig. 7, when the distance r from the toe reaches approximately 10 mm, the $C(r)$ value of each test series is different, i.e., the common

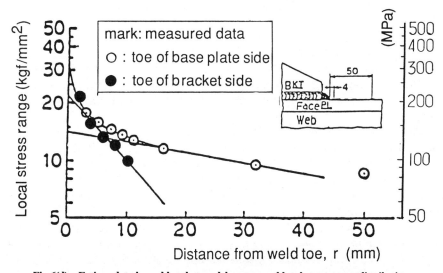

Fig 6(d)   Fatigue data by end bracket model – measured local stress range distribution

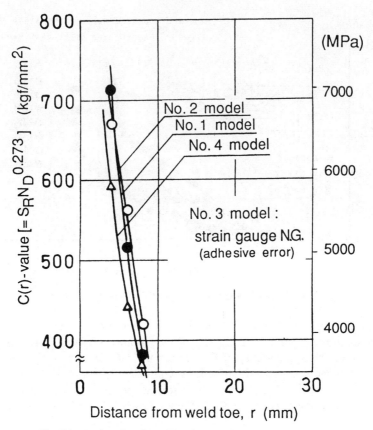

Fig 6(e)   Fatigue data by end bracket model – C(r) versus r diagram

fatigue strength diagram and equation (3) can be applied if $r \leqslant 5$ mm. Accordingly, when the stress range at $r = 5$ mm is the estimating stress, the common fatigue strength diagram can be used for all test series shown in this paper.

The S–N diagram given by the stress at $r = 5$ mm is shown in Fig. 8. The scatter of the data, and the applicability of equation (3) seemed to be satisfactory.

The above definition of estimating stress applies to the limited case where standard round fillet welding is performed by erecting a longitudinal stiffener with a plate thickness of 10–20 mm on a base plate 10–30 mm thick, which is the fatigue test data range used in this study. Also l_{cr} shown in Fig. 2, has been proposed as a critical crack configuration. It can, however, be supposed that crack propagation life N_P included in N_D tends to be reduced for the joint in which l_{cr} is larger. This is because of the effect of surface crack length on the crack growth in the direction of depth. In future, the precision study should be carried out by using a fracture mechanics approach on the influence of l_{cr} and other dimensions such as width of the specimen.

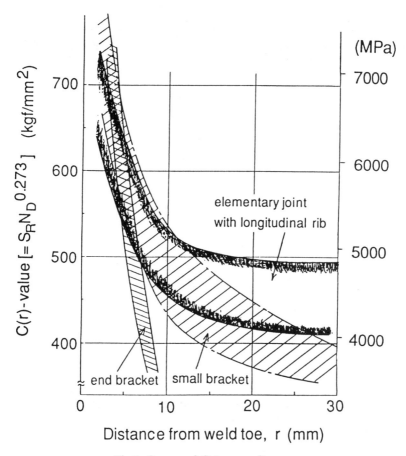

Fig 7   Superposed $C(r)$ versus r disgram

Conclusions

In order to acquire the estimation technique of fatigue strength which is superior in adaption to the direct strength calculation method, a fatigue test of a round fillet weld joint and the consideration for the estimating stress definition has been made. The results obtained are as follows.

(1) In the round fillet weld, fatigue crack initiation occurs at a considerably early stage compared with failure life ($N_i/N_f = 0.05$-0.1 in Fig. 1).

(2) In order to make the allowable crack configuration clear and significant in practical use the crack propagation life should be considered. A proposal is shown in Fig. 2 which considers the features in the crack growth behaviour.

(3) The local stress at 5 mm away from the toe of round fillet weld would be

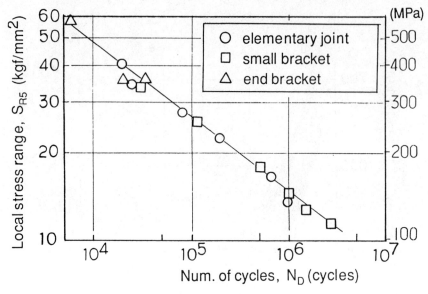

Fig 8 Fatigue strength diagram (S_{RS} versus N_d)

appropriate, for the practical fatigue strength evaluation given the plate thickness limitation in this study.

References

(1) HUTHER, M. and HENRY, J. (1991) Recommendation for hot-spot stress definition in welded joints, IIW-Doc. XIII–1416–91.
(2) KAWANO, H., KAWASAKI, T., SAKAI, D. *et al.* (1991) Some Considerations on reference stress definition in fatigue strength analysis. Trns. W. Jap. Soc. Nav. Archit., **83**, in Japanese.
(3) NIEMI, E. (1991) Recommendations concerning stress calculation for fatigue analysis of welded components, IIW-Doc. XV–763–91 and XIII–1426–91.
(4) PETERSCHAGEN, H., FRICKE, W., and MASSEL, T. (1991) Application of the local approach to the fatigue strength assessment of welded structures in ships, IIW-Doc. XIII–1409–91.
(5) The Shipbuilding Research Association of Japan, No. 202 Research Committee (1991) Study on Fatigue Design Method and Weld Quality of Offshore Structure, Report No. 395, in Japanese.
(6) The Shipbuilding Research Association of Japan, No. 207 Research Committee (1990)–(1992) and No. 216 Research Committee (1991)–(1993).

*K.-O. Edel**

Fracture Mechanics Analysis of the Residual Service Life of the Welded Main Girders of the Highway Bridges of Rüdersdorf

REFERENCE Edel, K.-O., **Fracture mechanics analysis of the residual service life of the welded main girders of the highway bridges of Rüdersdorf,** *Fatigue Design*, ESIS 16, (Edited by J. Solin, G. Marquis, A. Siljander, and S. Sipilä) 1993, Mechanical Engineering Publications, London, pp. 109–119.

ABSTRACT To assess fatigue cracks which start from welding defects of about 10 mm in height at the end of the inner cover plates, specimens were taken from bridges in Rüdersdorf. The fracture mechanics properties of the 55 year old steel and the stress intensity factors have been determined to analyse the instable and stable crack extension. The essential influence on the residual life is the increasing traffic density. The assessment of the critical crack size and crack growth shows a residual service life of some years.

Introduction

The 742 m long highway bridge over the Mühlenfließ is a well known bridge. It was built in 1936 using the steel St 52 (with $R_m \geqslant 510$ MPa, $R_e \geqslant 355$ MPa) which had then been recently developed. The girders are welded steel fabrications but the roadway is a concrete construction. The bridge became well known due to instable cracking in the main girders which occurred in January 1938; in two cases cracks destroyed the under flanges. The instable extension of the cracks was arrested in the web near to the upper flanges. At the end of the World War II the bridge was partially blown up, but was reconstructed after the war.

The western part of the Mühlenfließ bridge has been closed due to large fatigue cracks being found in the flanges of the main girders during the last few years (see Fig. 1). In addition to these cracks many other cracks and defects were present in the weldments. Using fracture mechanics, the requirement was to determine the residual life of the eastern part of the Mühlenfließ bridge and also that of the neighbouring Kalkgraben bridge. The ultimate objective of the investigations was to determine rebuilding priorities and to estimate the available time.

The assumed crack shape development

Practical experience shows that the greatest risk for bridges arises from fatigue cracks starting from welding defects at the end of the inner cover plates. In

* Fachhochschule Brandenburg, D-14770 Brandenburg (Havel), Germany.

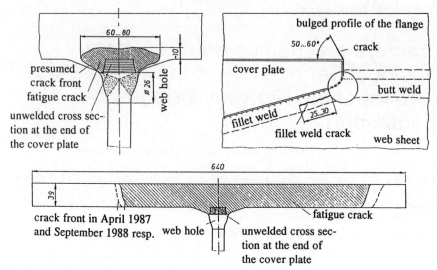

Fig 1   Fatigue cracks in the upper flange of the main girders of the Mühlenfliess bridge which start
from welding defects at the end of the inner cover plates

order to carry out a fracture mechanics analysis and assessment it is necessary
to make assumptions about the developing crack shape. In this work it has
been assumed that the crack shape develops in accordance with the experimental results of longitudinally stressed specimens with non-load-carrying
fillet welds by Maddox **(9)**. In this case the crack shape is given by the equation

$$2c = 2.58a + 6.71 \text{ mm} \tag{1}$$

Figure 2 shows the semi-elliptical fatigue cracks in the flanges of the
Mühlenfließ bridge which has a thickness of 39 mm; the special form of the
weld connection and special shape of the flange are considered. The crack
depth a is taken to be the vertical difference between the deepest crack front
point and the intersection of the crack front with the inner surface of the
flange. The fracture mechanics calculations show that the crack shape given by

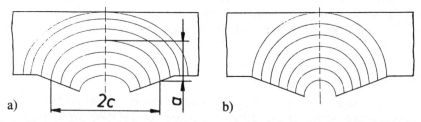

Fig 2   The assumed stages of development of the crack shapes in the flanges of the Mühlenfliess
bridge according to experimental experiences of (a) Maddox (9) and (b) JWES 2805 (8)

the relationship in the Japanese standard JWES 2805 **(8)** is not as suitable as those given by equation (1). Consequently, the crack shape given by JWES 2805 will not be considered in further detail here.

Following flange penetration, the fatigue cracks are assumed to be through-thickness cracks.

Stresses in the main girders

Based on stress calculations, the static stresses at the end of the cover plates are

$$\sigma_{stat} = 142 \text{ MPa}. \tag{2}$$

The theoretical maximum value of the dynamic bending stress range is

$$\Delta\sigma_{max} = 95 \text{ MPa} \tag{3}$$

The distribution of dynamic stresses caused by the effect of vehicles is characterized by the values given in Table 1. Assuming a number of 1200 vehicles a day, and applying a rain flow cycle identification procedure, gives 260 000 cycles per year. However, the closure of the western part of the Mühlenfließ bridge, and present and future increases in traffic, has resulted in a multiple of the cycles per year.

The dynamic bending stress $\sigma_{dyn\ max}$ is not known. For the calculations it is assumed that the dynamic bending stress is nearly equal to its range. Practical experience **(6)** suggests that the realistic value of the maximum dynamic stress is around 60 percent of its theoretical maximum value, i.e., the realistic value of the critical dynamic stress is 57 MPa.

Bridge stress measurements show frequencies of about 1–2 Hertz. The rise times of load or stress are given by

$$\Delta t = 20.2472/v \quad \text{(with } \Delta t \text{ in seconds and } v \text{ in km/h)}. \tag{4}$$

The loading rate for the critical case can be determined from the critical dynamic stress and the loading time given by equation (4).

Table 1 Classified values of the random bending stresses in the flanges near the supports of the Mühlenfließ bridge (14)

i	$\Delta\sigma_i/\Delta\sigma_{max}$	N_i	i	$\Delta\sigma_i/\Delta\sigma_{max}$	N_i
1	0.1	–	6	0.6	5500
2	0.2	810 000	7	0.7	1300
3	0.3	110 000	8	0.8	170
4	0.4	55 000	9	0.9	27
5	0.5	18 000	10	1.0	3

For the crack growth calculations the random values of the dynamic stress ranges are converted to an idealized equivalent constant amplitude stress range in accordance with the ECCS recommendations (12) as follows

$$\Delta\sigma_{eff} = \left(\frac{\Sigma(N_i \Delta\sigma_i^m)}{\Sigma N_i}\right)^{1/m} \tag{5}$$

Residual stress measurements show no decrease of the residual tensile stresses in the welded joint even though the structure has been in use for over 50 years. For the fracture mechanics calculations the maximum residual tensile stress is assumed to be the yield strength of the steel viz.

$$\sigma_R \approx 360 \text{ MPa} \tag{6}$$

With regard to the residual stresses near to the weld joint, it was assumed that the weld defect is free from residual tensile stresses.

Stress intensity factors

The stress intensity factor is the most appropriate parameter to describe the loading environment near to the crack tip. The stress intensity factors have been determined using several published solutions. The influence of the crack shape and the flange dimensions is dealt with using Isida's solution for the centre cracked strip (7) and Newman and Raju's solution for surface cracks (11). The distribution of residual stresses in the weld and the corresponding stress intensity factors are given by a through-the-thickness crack solution by Tada and Paris (15) (see (10)). The influence of the web on the stress intensity factor for a crack in the flange is dealt with a solution for crossing sheets given by Cartwright and Miller (1).

The superposition of the various effects characterized by the functions f_i is given by

$$K_I = \sigma\sqrt{(\pi a)} f \quad \text{with } f = \Pi f_i \tag{7}$$

The resulting function f, or calibration factor, is assumed to be the product of the various individual functions (6).

Fracture toughness of the girder steel

Specimens for the experimental determination of fracture toughness properties were taken from the projecting ends of the main girders of the Mühlenfließ and the Kalkgraben bridges. The valid K_{Ic} and K_{Id} values of 17 CT, 27 3PB, and 47 Charpy specimens are transferred to the bridge assuming that the critical temperature of the steel in the bridge is identical to the winter air temperatures. The distribution of the fracture toughness of the bridge girders is

Table 2 Quasi-static and dynamic fracture toughness of girder steel St52 under winter conditions

	K_{Ic}	K_{Id}
Mean value	3277.6 N/mm$^{3/2}$	1493.7 N/mm$^{3/2}$
5 percent lower quantile value	2864.6 N/mm$^{3/2}$	1009.1 N/mm$^{3/2}$
Loading rate	$\leqslant 85$ N/mm$^{3/2}$/s	$\approx 1.13 \times 10^7$ N/mm$^{3/2}$/s

determined from the fracture toughness probability and the winter time air temperature probability distribution as follows **(2)(4)**

$$P_{bridge}(K_{Ic, 0}; K_{Ic, u}) = \sum_{T=T_{c\,min}}^{T=T_{c\,max}} P_{steel}(K_{Ic, 0}; K_{Ic, u}; T)P_T(T_0; T_u) \tag{8}$$

The results of calculations using equation (8) are given in Table 2 in the form of mean and minimum fracture toughness values. This table also gives the appropriate strengthening rates.

The bridge girder is stressed by a high static load and an additional dynamic load. This is different to the fracture toughness specimens which are not pre-loaded. Given the loading time of the structure as Δt (see equation (4) and Fig. 3) the critical crack tip stress intensity rate in the specimen is given by

$$dK_I/dt \bigg|_c = \frac{K_{Ic}(dK_I/dt|_c)K_{Idyn\,max}}{\Delta t(K_{Istat} + K_{Idyn\,max})} \tag{9}$$

Within the limits shown in Fig. 4 the fracture toughness of the bridge girder steel, $K_{Ic}(dK_I/dt|_c)$, is assumed to be a linear function of log (dK_I/dt). By interpolation

$$K_{Ic}(dK_I/dt|_c) = K_{Id} + (K_{Ic} - K_{Id}) \frac{\log dK_I/dt|_{dyn} - \log dK_I/dt|_c}{\log dK_I/dt|_{dyn} - \log dK_I/dt|_{stat}} \tag{10}$$

Given the reference values in Table 2 and the static and dynamic stress intensity factors for the bridge girder crack it is possible to determine equations the stress intensity factor rate, $dK_I/dt|_c$, and the fracture toughness of the bridge steel, $K_{Ic}(dK_I/dt|_c)$ from equations (9) and (10)**(2)(4)**.

Fatigue crack growth rate

Specimens for the experimental determination of the fatigue crack growth rates were taken from the projecting ends of the Mühlenfließ bridge and the Kalkgraben bridge. The results are given in the form of the Paris equation

$$da/dN = C\Delta K^m \tag{11}$$

The upper bound line of the crack growth rate for the nineteen tested CT specimens has been taken two standard deviations above the regression line in

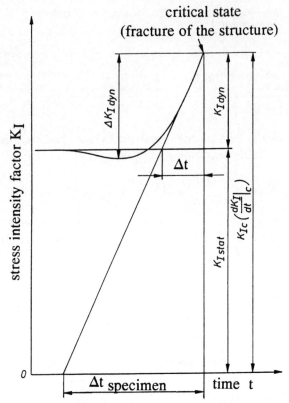

Fig 3 Schematic graph of the load increase due to the action of vehicles in a cracked bridge structure versus the time

accordance with the ECCS recommendations **(12)** as follows (see **(2)**)

$$\mathrm{d}a/\mathrm{d}N = 1.3C \exp{(2s_{\log \mathrm{d}a/\mathrm{d}N})}\Delta K^m \tag{12}$$

with

$$C = 1.534 \times 10^{-12} \ (\mathrm{mm/cycl})(\mathrm{N/mm}^{3/2})^{-2.761};$$
$$s_{\log \mathrm{d}a/\mathrm{d}N} = 0.213 \ (\text{with } \mathrm{d}a/\mathrm{d}N \text{ in mm/cycl});$$
$$m = 2.761.$$

The cyclic loading frequency in the crack growth tests was between 15 and 20 Hertz; however, the frequency in the bridge is lower. According to the experimental results of Yokobori and Sato **(16)** the difference in frequency results in a 30 percent increase in the crack growth rate. Such an increase has been shown in equation (12) by the factor 1.3. Effects of the R ratio were not significant in the crack growth tests. Line 8 in Fig. 5 is in accordance with equation (12).

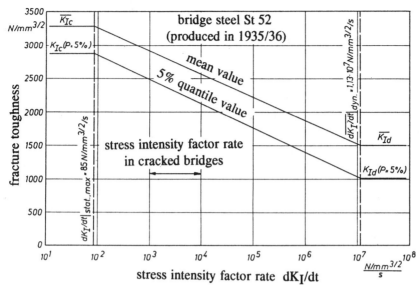

Fig 4 Idealized dependence of fracture toughness on stress intensity rate for the 55 year old bridge steel St52

The results of experiments with altogether 6 CCT specimens are given by lines 9 and 10 in Fig. 5. The tests consider a frequency of 5 Hertz and different R ratios. These results are of less interest because they give smaller crack growth rates than line 8.

Analysis and assessment of the critical crack size

The principles of small-scale yielding, linear–elastic fracture mechanics has been applied to describe the fracture behaviour of the 39 mm and 44 mm, respectively, thick flanges of the main girders.

Figure 6 compares the stress intensity factors in the structure with the material's fracture toughness under service conditions. The calculated critical crack sizes show a large scatter which is related to the dynamic stresses in the main girder. The vertical critical crack sizes a_c for the worst case are 21.0 mm and 17.8 mm for the Mühlenfließ and the Kalkgraben bridges, respectively, for a dynamic stress of 57 MPa and a vehicle speed of 60 km/h.

The service limit crack size, $a_{all\,max}$, is derived from the calculated critical crack size, a_c, by the relationship

$$S_{ac} = a_c/a_{all\,max} = 1.2 \tag{13}$$

The service limit crack sizes are, therefore, 17.5 mm and 14.8 mm for the Mühlenfließ and the Kalkgraben bridges, respectively.

The use of a relatively small safety factor of 1.2 is considered to be accept- able since the calculation is based upon a combination of unfavourable condi-

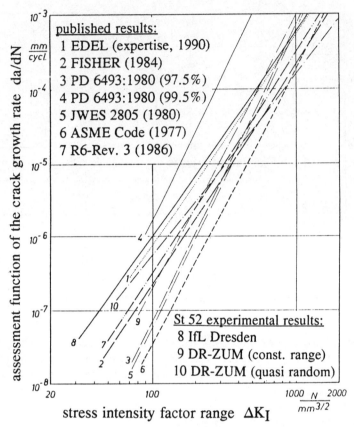

Fig 5 Published upper bound fatigue crack growth rate lines and the results from the present investigations

tions. In a fracture mechanics assessment of critical crack sizes in monobloc railway wheels, which was based upon the Monte Carlo simulation method, the safety factor used was 1.4 **(3)**. In this case the critical crack sizes represented a survival probability of 90 percent.

Assuming a critical temperature of $-25°C$, the fracture mechanics calculations and assessments lead to an unusable result. Based upon this temperature the critical crack sizes a_c in the welded structure have nearly the same size as the welding defects with depth of 10 mm according to non-destructive testings. In contrast to this theoretical result the bridges have survived for more than half a century, without fracture, with such welding defects.

Analysis and assessment of fatigue crack growth behaviour

The crack growth calculations for the deepest point of the crack front and penetration point of the crack front with the surface of the flange show larger

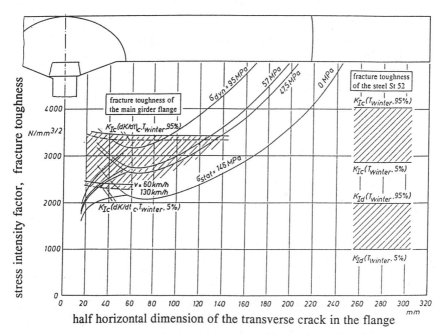

half horizontal dimension of the transverse crack in the flange

Fig 6 **Fracture toughness and stress intensity factors of the semi-elliptical crack (shape after Maddox) and the through-the-thickness crack in the flange of the Mühlenfliess bridge**

differences in cycles or in growing times for the crack shape according to JWES 2805 in comparison to the Maddox shape. Consequently the Maddox crack shapes are considered to be more realistic, and hence these have been used in the assessment of residual service life. In order to obtain a definitive crack growth life, the calculated lives at both crack front points are averaged.

The load spectrum was given by the values of the Table 1; the neglected 'non-damaging' small stress ranges and their influence on crack growth are not known. Consequently, the assessment of the crack growth from the initial crack size a to the service limit crack size $a_{\text{all max}}$, is given by the following relationship

$$a_{\text{all max}} - a = \Delta a_{\max} = S_{\Delta a}\,\Delta a_{\text{calc}}. \quad \text{with } S_{\Delta a} = 2 \tag{14}$$

The results of the fracture mechanics analysis and assessment are shown in Fig. 7 as representative fatigue crack growth lines. The intersection of these lines with the crack depth determined by non-destructive testing yields the residual service life for a given traffic density.

Comparing Fig. 7 with Fig. 2, it can be seen that the depth of the defect d_{defect} in the flange is not identical with the value a of the crack depth which is used in the fracture mechanics calculations. The relationship $d_{\text{defect}} > a$ exists as a result of the connection between the web and the flange of the girder.

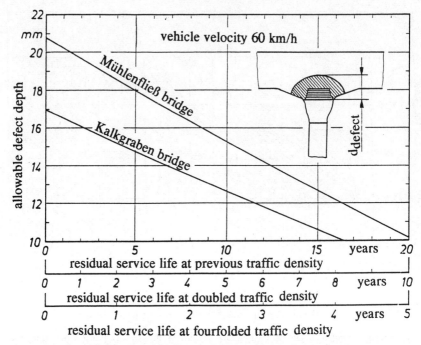

Fig 7   **Representative crack growth lines for the welding defects at the end of the inner cover plates for determining the allowable residual service life of the Rüdersdorf highway bridges**

Conclusions

A fracture mechanics analysis has been carried out on a certain type of weld defect contained in the main girders of the Rüdersdorf highway bridges. Measured and calculated stresses, experimentally-determined quasistatic and dynamic fracture toughnesses, and fatigue crack growth rates have been used to determine critical crack sizes and the residual life of the structure. The critical crack size depends on the velocity of the vehicles. The residual life of the structure depends on the density of traffic. The allowable crack size is assumed to be fraction of the critical crack size, but the maximum crack growth is assumed to be a multiple of its calculated increase.

The inspection period for the critical areas of the main girders should be shorter than the calculated residual life, due to the limited accuracy of the practicable non-destructive testing methods and the little experience with such a fracture mechanics assessment.

Only limited data can be generated from specimens removed from existing bridges; consideration must, therefore, be given to data from other studies, but the effect of the service environment of any specific structure must also be taken into account.

There is some uncertainty regarding safety factors in fracture mechanics assessments of cracked structures (13); this is only likely to be resolved by systematically collating data from calculations and practical experiences (5).

Acknowledgements

The author would like to thank Mr. Reuter, head of the Brandenburgisches Autobahnamt for permission to publish this paper. He would also like to acknowledge D. F. Cannon, British Rail Research, for his valuable and extensive contribution to this paper, and W. Spruch, Deutsche Reichsbahn, for his assistance in the technical realization.

References

(1) CARTWRIGHT, D. J. and MILLER, M. (1975) Stress intensity factors for a crack in a sheet with a partially debonded stiffener, *Int. J. Fracture*, **11**, 925–932.
(2) EDEL, K.-O. (1991) Die Ermittlung der bruchmechanischen Eigenschaften rißgeschädigter Brückenteile, *VDI-Bericht 902*, p. 313–342.
(3) EDEL, K.-O. (1987) *Untersuchung des Bruchverhaltens von Eisenbahnschienen und -vollrädern*, Dissertation, Technische Hochschule Otto von Guericke Magdeburg.
(4) EDEL, K.-O (1991) Bruchmechanische Analyse der Restnutzungsdauer der Hauptträger der Autobahnbrücken A1/Bw13 und A1/Bw15 bei Rüdersdorf, Expertise.
(5) EDEL, K.O. (1991) Sicherheitsbewertung des Rißverhaltens. Deutscher Verband für Materialforschung und -prüfung, Tagung "Werkstoffprüfung 1991", Vorträge der Tagung, p. 83–92.
(6) FISHER, J. W. (1984) *Fatigue and fracture in steel bridges – Case studies.* John Wiley.
(7) ISIDA, M. (1973) Analysis of stress intensity factors for the tension of a centrally cracked strip with stiffened edges, *Engng Fracture Mech.* **5**, 647–665.
(8) JWES 2805 (1980) Method of assessment for defects in fusion-welded joints with respect to brittle fracture. The Japan Welding Engineering Society.
(9) MADDOX, S. J. (1975) An analysis of fatigue cracks in filled welded joints, *Int. J. Fracture*, **11**, 221–243.
(10) MURAKAMI, Y. (1987) *Stress intensity factors handbook*, Pergamon Press, Oxford.
(11) NEWMAN, J. C. and RAJU, I. S. (1981) An empirical stress-intensity factor equation for the surface crack, *Engng Fracture Mech.*, **14**, 185–192.
(12) Recommendations for the fatigue design of steel structures, (1985) European Convention for Constructional Steelwork, Committee TC6 "Fatigue", ICOM 141.
(13) v. RONGEN, H. J. M. (1991) Review of safety factor treatments, International Institute of Welding, IIW–report X–1220–91.
(14) SCHMIEDER, SEIFERT and POHL (1987) Reko A1, Bw 13, Brücke Rüdersdorf, 2–4. Stufe der Nachrechnung. Statische Berechnung, p. 94–134.
(15) TADA, H. and PARIS, P. C. (1983) The stress intensity factor of a crack perpendicular to the weld bead, *Int. J. Fracture*, **21**, 279–284.
(16) YOKOBORI, T. and SATO, K. (1976) The effect of frequency on fatigue crack propagation rate and striation spacing in 2024–T3 aluminium alloy and SM-50 steel, *Engng Fracture Mech.*, **8**, 81–88.

J. M. Varona*, F. Gutiérrez-Solana*, J. J. González*, and I. Gorrochategui*.

Fatigue of Old Metallic Railroad Bridges

REFERENCE Varona, J. M., Gutiérrez-Solana, F., González, J. J., and Gorrochategui, I., **Fatigue of old metallic railroad bridges,** *Fatigue Design*, ESIS 16 (Edited by J. Solin, G. Marquis, A. Siljander, and S. Sipilä) 1993, Mechanical Engineering Publications, London, pp. 121–135.

ABSTRACT This work presents a study of the fatigue behaviour of materials taken from old metallic railroad bridges, as a prerequisite to determining their residual life. Following an appropriate method for the characterization in fatigue of these materials – carbon steels and puddle irons – their S–N Wöhler curves are obtained. These curves are analysed according to the European standards and recommendations on fatigue design of steel structures. The work is completed with the characterization of the behaviour of welded joints.

The most important conclusion is the relevant influence of the surface defects present in the material on its fatigue behaviour. The presence of these defects conditions the process of obtaining samples from the bridges to be analysed. These samples have to include the surface defects in order to be representative.

Introduction

The Spanish National Railway Network (RENFE) has researched the modernization of rolling stock and installations, in relation to the bigger loads and higher speeds that modern society demands. One of the most important parts of this study is the calculation of the remaining life of old metal bridges in relation to their load history – past, present, and future. In order to do this, the stresses caused by the loads and the fatigue caused by the constant repetition of these loads must be known in greater depth through the fatigue behaviour characterization of the bridge materials.

Sample extraction

The process of material characterization must be carried out in both the conventional chemical and metallographic analysis, weldability, and mechanical properties, and the less conventional fatigue. Samples were taken by mechanical means from perfectly identifiable structural elements. These pieces were conveniently replaced with new materials to guarantee safety. The samples were taken from parts of the structure subjected to low stresses in order to preserve the validity of the fatigue characterization. For purely economic reasons only a few samples were taken from each bridge. Then a much less than desired number of fatigue tests were carried out; this made it difficult to obtain a complete statistical treatment of the experimentally-obtained results.

* Department of Civil Engineering, University of Cantabria, Avenida de Los Castros, s/n 39005-Santander, Spain.

Standards

The structural analysis of these bridges requires a complete mechanical characterization of their materials under static and fatigue conditions. From this, the static capacity of the bridges for actual and future loads and the fatigue assessment of their parts which are subject to repeat fluctuations of stress can be determined. Using this, the residual lives of these bridges can be calculated.

The existing standards and recommendations for the fatigue analysis of metallic bridges (3)–(7) establish characteristic S–N curves for steels and, sometimes (7), for puddle irons. These curves vary according to a classification of constructional details of the parts of the bridges, welded or bolted. Nevertheless, they consider neither the mechanical characteristics nor the state of damage of the materials of the bridges. This lack of considerations, which is very important when the analysis is done on very old bridges to determine their residual life, as well as the existing differences on the S–N curves proposed by different standards, determined the necessity to obtain the Wöhler curves of the bridges' own materials analysed under their real state of defects. The corresponding curves for constructional details can be obtained from these, applying the reductions suggested by the different standards with reference to the base material behaviour, non-conditioned by welds or joints.

Objectives

According to previous considerations, the first part of this work was carried out using the material obtained from four railroad bridges from the Madrid–Hendaye line, some of them over 100 years old. Its objectives were to determine the S–N Wöhler curves and the crack propagation rate, $da/dN-\Delta K_I$, for the materials from these bridges. The method used was to apply fracture mechanics concepts on CT samples, and to correlate the obtained behaviour with the existing defects in the materials where cracks nucleate under fatigue conditions. Then a complete fractographical analysis by scanning electron microscopy was performed (8).

Secondly, after samples were taken from five different bridges on the Madrid–Seville line, the fatigue behaviour of their materials, which had been chemically, microstructurally, and mechanically characterized, was performed with two principal objectives; (a) to determine the influence of fatigue samples preparation as a function of the level of preservation of existing defects; and (b) to ascertain the effects of material characteristics, such as their microstructure, their level of precipitation or inclusions, and their mechanical behaviour on the fatigue resistance (9).

Thirdly, using some of the later samples, the work focussed on determining the fatigue behaviour of welded materials (10). This included the study of the corresponding morphology of fracture paths to justify the important reduction of the resistance to fatigue imposed by standards to the structural details with welded joints.

Experimental work

The samples taken from Madrid–Hendaye bridges were cut and machined to prepare the different specimens to be tested, these were: long flat specimens to determine fatigue curves, CT specimens to characterize crack propagation rate; and other conventional, tensile and Charpy type, specimens and coupons.

The samples from the bridges on the Madrid–Seville line were cut into two different groups for fatigue tests: (a) the external specimens, obtained from a longitudinal cut, that preserves the external border, or lateral surface, of the bridge elements; and (b) the internal specimens, obtained after two longitudinal cuts which do not preserve the border. In both of these groups the longitudinal direction of the samples corresponds to the load direction on the corresponding structural elements. Stress conditions are then reproduced and the intrinsic anisotropy of these rolled materials is avoided.

Materials

The materials from the Madrid–Hendaye bridges were steels and puddle irons. The steels, non alloyed, presented a ferritic–pearlitic microstructure according to their carbon content. The puddle irons, Fe–P alloys, presented a ferritic matrix with an important presence of slag impurities, oriented in the rolling direction, the size of which was up to 1 mm (Fig. 1).

As an example of the differences of these materials Table 1 shows the chemical composition and yield stress of the samples with extreme carbon contents for each of the alloy types present in the four bridges. Most of the samples presented a high content of sulphur and phosphorus, above the quantities admitted for structural steels in actual standards, as for example UNE 36–080–85. These materials can be set in different groups of steels and puddle irons, according to their strength. Taking yield stresses as an example, steels vary between 200 and 350 MPa, and puddle irons between 240 and 350 MPa.

Fig 1 Slag details in puddle iron

Table 1   Characteristics of materials (Madrid–Hendaye line)

Bridge	Material	Sample	Chemical composition (wt %)					Yield strength σ_y (MPa)
			C	Mn	Si	P	S	
1	Steel	6	0.133	0.64	0.12	0.064	0.051	293.3
		3	0.027	0.40	<0.03	0.035	0.029	197.7
2	Puddle	7	0.010	0.03	0.10	0.410	0.026	240.8
	iron	1	0.001	0.04	0.13	0.480	0.037	261.4
	Steel	4	0.088	0.42	<0.03	0.050	0.052	250.6
3		3	0.053	0.37	<0.03	0.041	0.058	300.7
	Puddle	8	0.008	<0.02	0.38	0.550	0.053	266.3
	iron	5	0.001	<0.02	0.39	0.600	0.067	254.1
4	Steel	2	0.206	0.52	<0.03	0.030	0.071	304.1
		6	0.036	0.46	<0.03	0.007	0.027	300.2

The materials from the Madrid–Seville bridges were, in all cases, low carbon ferritic–pearlitic steels, with high sulphur and phosphorus contents. Their strength varied between 250 and 360 MPa when measured as yield stress.

Experimental methodology

The fatigue behaviour was obtained following two complementary procedures. The first one, after a fracture mechanics approach and using the CT specimens, allowed the determination of the fatigue crack propagation rate of steel and puddle iron, represented by the da/dN–ΔK_I relations shown in Fig. 2. The second one performed the fatigue tests on the longitudinal flat specimens, choosing different levels of the amplitude of load variation. From both procedures a final Wöhler S–N curve is defined for each material.

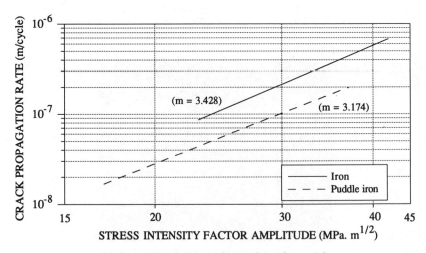

Fig 2   Fatigue crack growth rate of tested materials

From the samples of the Madrid–Seville bridges it was not possible to prepare CT specimens. Then, the S–N curves, $N = C' (\Delta\sigma)^{-k}$, should be determined using the values of k suggested by the different standards, the maximum and minimum being those corresponding to UIC (7), $k = 5$, and CECM (3), $k = 3$.

Specimen preparation

The CT specimens were machined preserving the complete thickness of the samples obtained from the bridges. Other dimensions were chosen to obtain a geometry previously tested and calibrated. Their initial flaw was machined in the transversal plane of the sample.

The cut surfaces of the long flat specimens were treated, rounded, and polished to avoid the initiation of fatigue cracks. Both ends of these specimens were covered with steel liners fixed by an epoxi-resin mortar, avoiding damage by the grips of the testing machine (11).

The welded specimens were obtained from internal samples that were transversally cut to form a 30 degree weld edge, and butt welded using an OK 4800 (UNE E 455 B20) electrode 3.25 mm in diameter. Finally they were ground and welded again with the same type of electrode (2).

Test performance

The fatigue crack propagation rate was measured on the CT specimens, after a compliance determination. The compliance was determined from the continuous measurements obtained from a COD dynamic extensometer and the load cell of the testing machine at the maximum and the minimum of the load level, P.

$$\text{Comp} = \frac{\text{COD}_{max} - \text{COD}_{min}}{P_{max} - P_{min}} \tag{1}$$

The compliance values were correlated with the obtained crack lengths, measured optically at the different positions, where the tests were stopped. Once this correlation was obtained all the compliance determinations, which were sequentially obtained, were indirect measurements of crack length. So, crack propagation rate can be determined between each two measures of crack length, as a function of the mean value of the corresponding stress intensity factor, ΔK_I. The ΔK_I values were obtained directly from the crack length, a, the specimen geometry, and the load amplitude, following a calibrated relationship known for the specimens used.

Crack rates were determined between crack lengths separated enough to avoid microstructural discontinuity. Figure 3 shows the results obtained for the steel characterized.
So, the steel fatigue cracks grow following the equation

$$da/dN = 1.85 \times 10^{-12}(\Delta K_I)^{3.43} \tag{2}$$

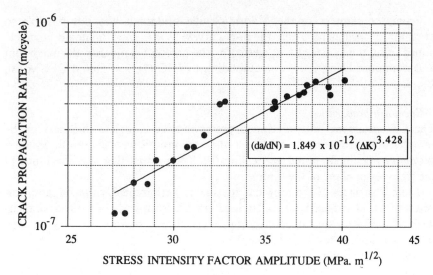

Fig 3   Paris law for steel

In the same way, puddle irons were characterized, offering the following crack propagation rate

$$da/dN = 2.08 \times 10^{-12}(\Delta K_3)^{3.17} \tag{3}$$

The long flat specimens from all the bridges were tested in fatigue, each one at a different stress amplitude variation, up to 90 percent of σ_y, in a first step to 2×10^6 cycles or rupture. The specimens that achieved those cycles were tested again in a second fatigue step at higher stress amplitude.

Tables 2 and 3 show the test parameters of the specimens from bridge 1 steels and bridge 2 puddle irons respectively, as well as the cycles at rupture, N, at the corresponding step of $\Delta\sigma$.

Figure 4 shows the results listed in Table 2 in a bilogarithmic representation, log N–log $\Delta\sigma$, in order to correlate them with the S–N curves suggested by the existing standards and recommendations for the fatigue analysis of metallic bridges.

Table 2   Fatigue tests. Bridge 1 (steel)

Sample	Size (mm)	Yield strength σ_y (MPa)	$\Delta\sigma$ (MPa) 1st part	$\Delta\sigma$ (MPa) 2nd part	ΔP (kN) 1st part	ΔP (kN) 2nd part	Cycles 1st part	Cycles 2nd part
1	9.7 × 27.8	201.0	176.6	—	47.7	—	1 365 679	—
2	10.0 × 27.0	294.3	255.1	—	68.9	—	206 260	—
3	11.3 × 25.3	196.2	157.0	—	44.9	—	672 168	—
4	11.5 × 25.0	201.1	137.4	186.4	39.5	53.6	2 000 000	72 759
5	10.7 × 25.0	294.3	255.1	274.7	68.2	73.5	2 000 000	1 543 196
6	10.7 × 26.3	294.3	235.5	—	66.3	—	1 683 054	—
7	12.3 × 26.5	294.3	215.8	264.9	70.3	86.3	2 000 000	463 188
8	12.0 × 25.0	294.3	196.2	255.1	58.9	76.5	2 000 000	721 757

Table 3   Fatigue tests. Bridge 2 (puddle iron)

Sample	Size (mm)	Yield strength σ_y (MPa)	$\Delta\sigma$ (MPa)		ΔP (kN)		Cycles	
			1st part	2nd part	1st part	2nd part	1st part	2nd part
1	12.0 × 28.2	260.0	215.8	—	73.0	—	25 971	—
2	12.0 × 24.0	240.3	176.6	206.0	50.9	59.3	2 000 000	618 457
4	9.5 × 23.5	245.3	215.8	—	48.2	—	156 449	—
6	10.0 × 24.5	240.3	196.2	—	48.1	—	263 781	—
7	10.0 × 24.0	240.3	196.2	—	47.1	—	399 038	—
8	10.5 × 22.3	235.4	176.6	—	41.3	—	71 957	—
9	10.0 × 21.5	264.9	196.2	225.6	42.2	48.5	2 000 000	742 417
11	10.5 × 22.0	279.6	235.4	—	54.4	—	435 222	—

The fractographic analysis of fracture surface carried out after each fatigue test allowed the determination of representative results when the specimens were tested in two steps. With reference to bridge 1 (Table 2 and Fig. 4), specimens 5, 7, and 8 showed only one fatigue crack growth area associated to the second fatigue step, their first one being under their threshold conditions. On the other hand, the fracture surface of specimen 4 presented two clearly differenciated fatigue growth areas. This explains the very low response of the specimen on its second fatigue step, point 4′ in Fig. 4, associated with the fatigue crack obtained at the first and not the original defects.

Fatigue characterization

The crack propagation rate determinations were used to obtain the slope of the Paris law, m in $da/dN = C(\Delta K_I)^m$. This slope has been considered as the k value that defines the slope corresponding to the intrinsic S–N behaviour of each specimen tested in fatigue, similar to the $N = C'(\Delta\sigma)^{-k}$ suggested by the standards. The obtained m values, 3.428 for steel and 3.174 for puddle irons, agree with the k values used by the standards, varying between 3 and 5, and

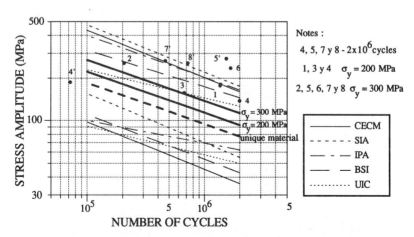

Fig 4   Fatigue tests and S–N curves – bridge 1

with previous values obtained by other research works for ferritic–pearlitic steels, that show a variation between 3 and 4.2 **(12)**.

For each bridge, or each group of specimens according to their material characteristics, a statistical methodology was used to define the corresponding S–N curve to be used in the corresponding structural analysis.

From the result obtained, $\Delta\sigma-N$, on the fatigue test of each specimen, and using the slope defined by k, the stress amplitude, $\Delta\sigma_{2\times10^6}$, corresponding to a number of cycles at rupture of 2×10^6 was determined.

In each group of analysis, the mean value and the standard deviation of these $\Delta\sigma_{2\times10^6}$ were determined, and the $\Delta\sigma_c$ value was calculated as the mean minus two times the standard deviation. The $\Delta\sigma_c$ value, considering a normal distribution, represents a confidence level of 97.5 percent.

The S–N curve representative of the corresponding group of analysis was defined as a straight line passing through the point $\Delta\sigma_c - 2 \times 10^6$ cycles, and with the slope $-1/k$ in a double logarithmic $\log N - \log \Delta\sigma$ representation. As an example Fig. 4 shows the S–N curves corresponding to the 200 MPa and 300 MPa yield strength steels from bridge 1, considered separately and as one group only.

Fractography

On each specimen tested, macrographs of its fracture surface were taken, as described above (Figs 5 and 6). Also a SEM fractographical analysis was done on each, to determine the morphology of fracture path and the defect where crack was initiated (Fig. 7). This analysis, when used on the welded specimens tested in fatigue, showed that cracks propagate, in all the cases, from slags formed at welding (Figs 8 and 9) **(10)**.

Results

The obtained results are classified according to the different parts of this work.

Fig 5   Steel fracture

Fig 6   Puddle iron fracture

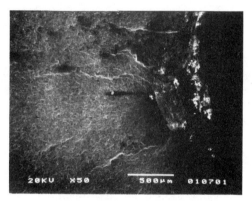

Fig 7   Initial defect in steel

Fig 8   Welded specimen fracture

Fig 9   Welding slag at crack initiation

First part

Following the methodology previously explained to obtain the representative fatigue S–N curves of the tested specimens, the different groups of materials obtained from the four bridges of the Madrid–Hendaye line have been characterized.

Similar analyses to the one shown for bridge 1 in Fig. 4 have been carried out for the other three bridges. Table 4 summarizes the $\Delta\sigma_c$ values obtained from which the S–N curves can be derived **(8)**.

Second part

The same methodology was followed for the two different groups of long flat fatigue specimens used in this part of the work, both external and internal. The k values used have been the maximum and minimum suggested by the standards i.e. three from CECM and five from UIC. The material has been considered as the same for all the samples taken from the five different bridges of the Madrid–Seville line **(9)**.

Tables 5 and 6 summarize the statistical values obtained for both external and internal specimens, considering UIC or CECM k values. Figure 10 shows the S–N curves obtained using the k value of UIC that offered a lower scattering.

Table 4   Fatigue characterization of Madrid–Hendaye line materials

Bridge	Material	σ_y (MPa)	$\Delta\sigma_c$ (MPa)
1	Steel	300	112.5
		200	92.7
		All (200–300)	77.2
2	Puddle iron	All (240–280)	28.3
3	Steel	All (250–300)	60.4
	Puddle iron	All (250–350)	24.5
4	Steel	All (260–350)	100.9

Table 5   Statistical results of external specimens (MPa)

Statistical values	CECM ($k = 3$)	UIC ($k = 5$)
$\overline{\Delta\sigma}_{2 \times 10^6}$	162.6	190.5
Standard deviation (s)	31.2	21.4
$\Delta\sigma_c$	100.2	147.7

Table 6   Statistical results of internal specimens (MPa)

Statistical values	CECM ($k = 3$)	UIC ($k = 5$)
$\overline{\Delta\sigma}_{2 \times 10^6}$	211.8	236.4
Standard deviation (s)	38.5	25.3
$\Delta\sigma_c$	134.8	185.8

Third part

After testing the welded specimens, the obtained data have been analysed in the same way **(10)**. The statistical values and the obtained $\Delta\sigma_c$ are summarized in Table 7. Figure 11 represents the S–N curves defined after this analysis for the welding detail defined in this work as a reference.

Discussion

The discussion of the obtained S–N curves is organized in correlation with the results.

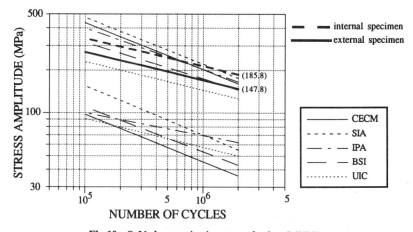

Fig 10   S–N characterization curves for $k = 5$ (UIC)

Table 7   **Statistical results of welded specimens
(MPa)**

Statistical values	CECM ($k = 3$)	UIC ($k = 5$)
$\overline{\Delta\sigma}_{2 \times 10^6}$	114.5	160.8
Standard deviation (s)	25.2	19.0
$\Delta\sigma_c$	64.1	122.8

First part

Figure 4 shows a good correlation between the determined $\Delta\sigma_c$ values and the yield stress of steels from bridge 1. This apparent correlation could not be extensively studied as the other bridges did not present a set of specimens of material with the same mechanical characteristics, and so they could not be grouped (Table 4).

As the existing standards and recommendations do not differentiate the fatigue behaviour of materials as their mechanical characteristics change, all the results obtained in the first part of this work have been analysed in two groups: steel and puddle iron. Table 8 shows the corresponding statistics and $\Delta\sigma_c$ values. It can be seen that both materials present a very similar conventional mechanical behaviour, not only in mean values but also in scattering, defining very close design values which maintain the same level of confidence.

Nevertheless, fatigue behaviour is very different for steels and puddle irons. Steel presents higher mean values of fatigue resistance and lower scattering, and S–N design curves establish a much better fatigue behaviour for steels than for puddle irons, 73.8 MPa–24.1 MPa in $\Delta\sigma_c$ values.

Second part

The two groups of specimens tested, internally or externally cut, present important differences on their determined S–N fatigue curves. Internal speci-

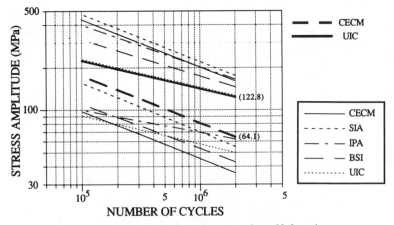

Fig 11   **S–N fatigue characterization curves for welded specimens**

Table 8   Statistical analysis of all fatigue results (Madrid–Hendaye)

Material	$\overline{\sigma_y}$	s	$\overline{\sigma_y} - 2s$	$\overline{\Delta\sigma_{2 \times 10^6}}$	$\Delta\sigma_c$
Iron	270.8	34.9	201.0	156.8	73.8
Puddle iron	263.2	30.3	202.5	88.3	24.1

mens were more resistant to fatigue cracking, having a mean $\Delta\sigma_{2 \times 10^6}$ value 35 percent (CECM analysis) or 25 percent (UIC analysis) higher than that corresponding to external specimens. As the scattering observed was the same for both types of specimens, $\Delta\sigma_c$ values keep similar differences.

In order to determine if these different behaviours were affected by the variation of metallurgical characteristics of the steels, the fatigue behaviour of each specimen, represented by the $\Delta\sigma_{2 \times 10^6}$ value, was compared with different material variables such as the level of inclusions and precipitates measured indirectly through the sulphur and phosphorus contents, the ferrite–pearlite relative presence related basically to the carbon content, and the strength behaviour, characterized by the yield and ultimate stresses. None of these comparisons offered a correlation which explained or justified the observed fatigue differences. As an example Fig. 12 shows the comparison between $\Delta\sigma_{2 \times 10^6}$ and sulphur content for both types of specimens, where no correlation is observed.

The differences in fatigue behaviour are exclusively associated to the preparation procedures of specimens by the way that they preserve the real surface defects of the structural elements of the bridges. The complementary fractographical analysis showed up the important influence of the lateral defects present on the external specimens and avoided when preparing the internal ones.

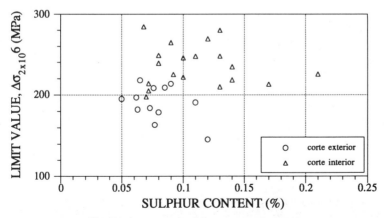

Fig 12 $\Delta\sigma_{2 \times 10^6}$ and sulphur content relationship

Third part

The analysis of the fatigue behaviour of welded specimens, and its comparison with that for base materials, shows the fatigue resistance reduction associated to the presence of welded joints with reference to non-jointed elements. The $\Delta\sigma_c$ values obtained, 64.1 MPa (CECM) and 122.7 MPa (UIC), represent 48 percent and 66 percent, respectively, of the corresponding values for base material when tested with internal specimens, from which welded samples were prepared (10).

As in base material specimens, the UIC approach offered lower scattering on the results obtained. For this approach, the observed reduction in fatigue resistance agrees with the coefficient of reduction proposed by this standard for type III and IV welding details, 0.65 and 0.50, respectively.

Conclusions

The study of safety structural assessment of old metallic bridges under fatigue conditions should be based on the previous characterization of the fatigue behaviour of their own materials. The work done has shown for the cases studied that the mean values representative of the fatigue resistance of these materials, 156.8 MPa for the steel and 88.3 MPa for puddle irons, are close to those proposed by existing standards.

Nevertheless, the important variation of the shape and dimensions of the present defects on these materials, where fatigue cracks initiate, causes a big scattering in fatigue resistance and then determines very low design values, $\Delta\sigma_c$, to preserve an appropriated level of confidence. These $\Delta\sigma_c$ values define S–N design curves for the different materials tested which represent a lower resistance to fatigue by comparison with the curves proposed by standards.

The importance of the present defects establishes the conditions to obtain samples and specimens. The defects should be preserved at maximum to improve their representation. Then, the standards should define the appropriate processes to obtain samples, and also to prepare specimens to be tested, from the bridges to be analysed to guarantee representation.

Usual welding procedures applied in structural elements produces important slags on welded areas, where fatigue cracks initiate. The fatigue resistance is much lower at welded joints by comparison with that corresponding to base materials. The obtained values of this reduction agree with those established by the different standards and recommendations.

For the methodology used here, the results obtained offered a lower scattering when UIC recommendations were followed.

References

(1) VARONA, J. M., GUTIÉRREZ-SOLANA, F., and GONZÁLEZ, J. J. (1989) Caracterización convencional y en fatiga de material de puentes metálicos de ferrocarril. Estudio de cuatro puentes de la Línea Madrid-Hendaya, Departamento de Ciencias e Ingeniería de la Tierra, el Terreno y los Materiales, Universidad de Cantabria, Final reports.

(2) VARONA, J. M., GUTIÉRREZ-SOLANA, F., and GONZÁLEZ, J. J. (1989) Caracterización convencional y en fatiga de material de puentes metálicos de ferrocarril. Estudio de cinco puentes de la Línea Madrid-Sevilla, Departamento de Ciencias e Ingeniería de la Tierra, el Terreno y los Materiales, Universidad de Cantabria, Final report.

(3) Recommandations pour la vérification à la fatigue des structures en acier. Recommandations de la Convention Européenne de la Construction Métallique (CECM), (1985) p. 33.

(4) SN 555 161. Constructions métalliques. Société suisse des Ingénieurs et des Architectes (SIA), (1979).

(5) Propuesta sobre la comprobación a Fatiga. Instrucción Puentes de Acero (IPA), (1982) p. 20.

(6) BS 5400. (1980) Steel, concrete and composite bridges, British Standards Institution (BSI), p. 56.

(7) UIC 779-1 R. Recommandations pour la détermination de la capacité portante des structures métalliques existantes, Union International des Chemins de fer (UIC), (1986).

(8) VARONA, J. M., HERNÁNDEZ, A., GORROCHATEGUI, I., GUTIÉRREZ-SOLANA, F., and GONZÁLEZ, J. J. (1989) Caracterización en fatiga de material de puentes metálicos antiguos de ferrocarril, *Anales de Mecánica de la Fractura*, **6**, 231–238.

(9) VARONA, J. M., GUTIÉRREZ-SOLANA, F., ALVAREZ, J. A., and GONZÁLEZ, J. J. (1990) Comportamiento en fatiga de acero estructural de antiguos puentes de ferrocarril, *Anales de Mecánica de la Fractura*, **7**, 63–70.

(10) VARONA, J. M., GUTIÉRREZ-SOLANA, F., GONZÁLEZ, J. J., SÁNCHEZ, L., and ALVAREZ, J. A. (1991) Comportamiento en fatiga de uniones soldadas y su influencia en la resistencia a fatiga de puentes metálicos, *Anales de Mecánica de la Fractura* **8**, 77–83.

(11) VARONA, J. M., GUTIÉRREZ-SOLANA, F., GONZÁLEZ, J. J. (1989) Optimización del sistema de anclaje para ensayos de fatiga de armaduras activas pretensadas (alambres y cordones), *Hormigón y Acero*, **172**, 99–104.

(12) ROLFE, S. T. and BARSOM, J. M. (1977) Fracture and fatigue control in structures, *Applications of Fracture Mechanics*, Prentice-Hall.

*A. Bignonnet**

Towards a Better Fatigue Strength of Welded Steel Structures

REFERENCE Bignonnet, A., **Towards a better fatigue strength of welded steel structures,** *Fatigue Design*, ESIS 16 (Edited by J. Solin, G. Marquis, A. Siljander, and S. Sipilä) 1993, Mechanical Engineering Publications, London, pp. 137–155.

ABSTRACT This paper reviews different postweld improvement techniques and their potential for improving fatigue strength. All these techniques are used to delay crack initiation for as long as possible. The most effective method consists of decreasing local stress in welded areas, either through geometric modification or by inducing favourable residual stresses in sensitive areas. The techniques discussed include the use of special electrodes and improved welding procedures, TIG or plasma dressing of the weld toe, weld toe grinding, and hammer and shot peening. The effects of various parameters, especially that of base metal mechanical properties, on fatigue strength are examined.

Introduction

The fatigue strength of welded steel structures is one of the main concerns of engineering bureaux, fabrication yards, and certification organizations. The effort to optimize structures is hampered by technical limitations; in the case of welded structures, the fatigue strength of the welded joints is often one of the main constraints. The welding operation itself produces geometric discontinuities which are prime candidates for initiation of cracks, either internal (at the weld root where there is lack of penetration) or external at the weld toe (undercut, slag inclusion).

Because of these defects, the fatigue crack initiation period is very short and the life of the structure is dominated by the propagation behaviour of cracks. Crack propagation characteristics caused by fatigue in structural steel are independent of mechanical properties and metallurgical structures. Thus, using a high strength structural steel instead of a normal grade steel will not extend the structure's lifetime unless measures have been taken to eliminate possible weld defects.

In order to improve fatigue strength or to safeguard against fatigue-induced rupture, the internal condition of the weld should be checked to ensure that defect and lack of penetration will not cause cracking. Thus the problem of fatigue reported at the weld toe can be treated using an appropriate finishing technique (general information can be found in references **(3)(14)(15)(17)**). Finishing enables an initiation period to be added to the structure's lifetime and provides protection against the possibility of fatigue-induced cracks appearing.

The ability of steels to withstand crack initiation increases with yield strength; as a result, good weld finishing enables high strength structural steels

* PSA.DRAS, Chemin de la Malmaison, 91570 Bievres, France.

to be used in structures which then benefit from the other advantages of such steels.

In summary, various methods can be used to achieve this goal:

– improving the welding process itself; use of special electrodes to obtain a gradual transition at the weld toe;
– weld toe dressing to provide a smooth weld bead-based plate junction without defects;
– grinding of the weld toe to eliminate defects and improve the weld bead-base plate joint;
– weld toe hammer peening to modify the defect geometry and surrounds defects with a field of residual compressive stress;
– shot peening to produce a field of residual compressive stress without modifying geometry.

Welding process improvement

Electrodes with improved fluidity

This method for improving the fatigue strength of welded joints does not involve any additional weld toe finishing. Choosing an electrode with a suitable coating enables a smooth geometric transition at the weld to be obtained as a result of improved filter metal fluidity.

Several studies (see for example references **(18)(20)** have shown that the use of special electrodes with improved fluidity resulted in a significant increase in the fatigue strength for instance, with 15 mm thick cruciform welds. For steels with a yield strength of 400, 600, and 800 MPa, the authors **(20)** obtained fatigue strength improvements of 48, 85, and 75 percent, respectively, for welding in the horizontal position.

The results obtained for joints welded with conventional electrodes and with improved fluidity electrodes show a clear difference in fatigue strength; this improvement can be attributed to a marked decrease in the stress concentration at the weld toe, Fig. 1.

The works cited in this section deal with horizontal welding, which is obviously favourable to a geometrically smooth weld toe. Nonetheless, this technique is of economic interest and additional studies, especially for welding in other positions, would be useful.

Improved weld profile

Improved welding procedures were devised **(1)(9)** to ensure optimum welded joint fatigue strength. The properties of this welding procedure can be summarized in terms of geometry, practice, and metallurgy.

With regards to geometry, the overall shape of the weld is as smooth as possible around the weld toe (see Fig. 2) so as to obtain an improved stress

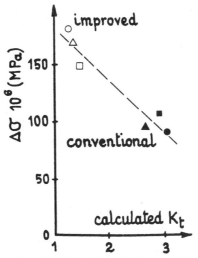

Fig. 1   Improvement of fatigue strength using special electrodes (20)

distribution at the weld toe and to reduce the overall stress concentration factor.

In practice, the weld toe run is performed after the root passes (see Fig. 2). This enables accurate positioning of the weld toe; the pass is performed without covering previous passes. Experience has shown that this procedure produces a better weld toe and significantly reduces the risk of shape imperfections.

Local elastic stress is limited in the improved profiles studied, whereas it is not limited in conventional profiles, due to discontinuity at the weld toe.

With regard to metallurgy, the toe pass is softened and stress relieved by heat treatment from the filling runs, since the toe pass is not the last pass to be performed.

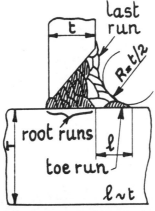

Fig 2   Improved weld profile (4)

Figure 3 **(4)** compares the fatigue strength results obtained using as-welded E460 steel test specimens made as per the improved welding procedure with data obtained from literature on 40 mm thick 'T' or cruciform joints of E355 steel welded using a conventional procedure.

The good results obtained using the improved weld profile can be attributed to the lower overall and local stress concentrations and reduced (or non-existent) residual tensile stress in the weld toe area. Crack initiation detected using the above-mentioned method corresponds to 20 or 30 percent of fatigue life. Furthermore, the benefits of weld improvement techniques remain even under seawater conditions when cathodic protection is used.

Inspection

Weld bead inspection is primarily visual. From the point of view of overall geometry, the weld bead should gradually blend into the sheet metal so that the overall stress concentration is limited. Sharp geometric discontinuities, as shown in Fig. 4, should be avoided for the following reasons:

(a) avoid a straight profile with too sharp an angle at the weld toe;
(b) avoid excessive concavity resulting in an angle recessed in the weld bead, and/or lack of filler metal in the weld toe area;
(c) respect the indications given in Fig. 2 and ensure that adequate over-thickness is provided in the weld toe area.

The inspector should use a template when checking the weld.

In order to achieve an optimum joint between the weld bead (local geometry) and the sheet metal (see Fig. 5), the weld toe pass can be laid down

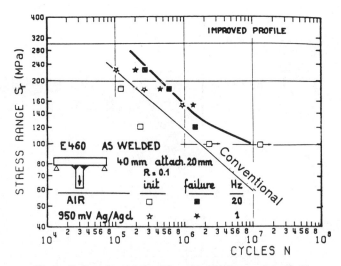

Fig 3   **Fatigue strength for fillet welds with improved profile (4)**

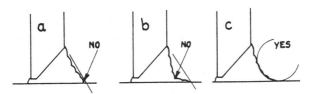

Fig 4   Control of the global geometry for improved weld profile

after the penetration passes, as shown in Fig. 2. Thus, the weld toe pass is not laid over the previous passes and spreads more freely over the sheet metal. Local geometry is significantly improved and the risk of undercuts is reduced. Inspection can be performed using a cut disk (ϕ20, height 17 mm) whose angle must touch the weld toe, as shown in Fig. 5. This inspection replaces the dime test which, in some cases, has been shown to be unsatisfactory. Cap passes should overlap by at least half a pass width so as to prevent undercuts between passes.

TIG dressing of weld toes

Finishing using TIG welding consists of dressing the weld toe to prevent fatigue cracks in this zone using a TIG torch. Stress concentration is reduced by smoothing the weld profile at the connection and increasing hardness in the treated area.

Operating procedure

The main reason for the improved fatigue strength of joints welded using TIG finishing is that the stress concentration at the weld toe is considerably reduced by the smoothing of the connection profiles and that defects are eliminated by dressing. The best results are obtained when the finishing pass is performed without using filler metal. Figure 6 shows a weld bead profile after double TIG dressing.

Pre-treatment

Scale deposits on the surface to be treated may cause small notches or undercuts along the remelted zone. For this reason, the surface should be free of scale, slag, and other impurities before TIG dressing.

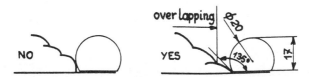

Fig 5   Control of the toe geometry

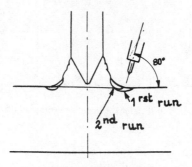

Fig 6   Two run TIG dressing (4)

TIG welding equipment

The heat supply required for TIG dressing is 10–20 kJ/cm. A high-capacity station enables the effect of the finishing treatment to be stabilized by increasing the heat input and improving output by increasing the speed of the operation.

If the tip of the electrode is damaged by oxidation during use, it should be reground or replaced.

Shielding gas

If the argon flowrate during TIG welding is inadequate, the arc becomes unstable and defects such as holes in the weld bead or electrode oxidation may occur. Given that numerous parameters such as nozzle dimensions, welding conditions, and welding environment (workshop or site) affects the gas flow rate required to prevent defects from occurring, it is generally considered that an initial TIG finishing test should be performed to determine the optimal flow rate. Some examples of suitable flow rate include 7 L/min. for a 12.7 mm diameter nozzle and 10 L/min. for a 15.8 mm diameter nozzle.

Dressing performance conditions

Although dressing conditions vary with welding position, the main goal of obtaining uniform weld beads free from blisters remains. Several sample welding conditions (23) are shown in Fig. 7. Current and speed ranges are also given.

The effect of TIG dressing depends to a large extent on the specific characteristics of each piece of welding equipment and on the initial weld bead profile; it is preferable to carry out an initial test in order to determine the optimal combination of dressing parameters and electrode orientation. Generally speaking, for convex weld beads, the electrode should be directed more towards the base metal. For flatter profiles, the electrode should be directed closer to the connection.

Two-run TIG dressing

TIG dressing increases hardness in the heat affected zone (HAZ) close to the dressed area, especially for higher carbon steels. Hardness of more than

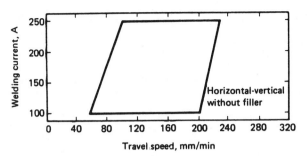

Fig 7   Example of TIG dressing parameter (23)

400 HV has been obtained. For low-carbon steels (approximately Ceq 0.40 percent), no risk of this type of problem exists. A second pass, 3–4 mm from the first (Fig. 6), causes annealing of the martensitic area formed by the first pass and, if necessary, reduces hardness at the weld toe to an acceptable level. This second pass enables a more uniform profile to be obtained.

TIG dressing technical parameters
Table 1 gives different experimental conditions used for TIG dressing.

Table 1   Examples of TIG dressing conditions

Reference		**(2)**	**(4)**	**(12)**
Steel		E 355	E 460	E 355/460/690
Nb run		2	2	1
Position		Vert. down	Vert. up	
Tungsten-electrode	mm	ϕ 3.2	ϕ 3.2	ϕ 4
Preheat	°C	100–150	75	
Shielding gas		Ar, 7 L/min.	Ar, 10 L/min.	Ar, 10 L/min.
O nozzle	mm	11		14
Voltage	V	12.5	13	17
Current	A	210	160	225
Speed	cm/mn	12	9	8
Heat input	KJ/cm	13	14	28

Fatigue strength

The results obtained by various authors, detailed in reference (3), enable some conclusions with regard to TIG dressing to be drawn. The results of tests performed on thick plates (2)(4)(6)(13) show that improvement due to TIG dressing remain fairly low in the high stress range ($N < 10^6$ cycles). On the other hand, for long lives, the improvement is very clear (Fig. 10). Improvements of 100 percent can be obtained for fillet welds of high strength structural steel.

Effect of joint geometry
Conventional fatigue strength improvement ($\Delta\sigma$ over 2×10^6 cycles) is much more significant for fillet welds than for butt welds. This is caused by the

marked reduction of the stress concentration coefficient in fillet welds after TIG dressing.

Effect of mechanical properties
The higher the yield strength of the material, the greater the improvement obtained using TIG dressing.

– For butt welds, fatigue strength improvement is approximately 10 percent for a steel with a yield strength of 250 MPa; it is 60–80 percent for steels with a yield strength greater than or equal to 400 MPa **(3)**.
– For fillet welds, the improvement is 20–50 percent for steels with a low yield strength (<400 MPa). For steels with higher yield strengths (>400 MPa), improvement exceeds 60 per cent and increases with the yield strength, until values of more than 100 percent are reached (Fig. 8).

Effect of mean stress
Minner and Seeger **(24)** studied the effect of the stress ratio R (-3, -1, 0, 0.4, 0.7) on high strength structural steel welded joints. Simon and Bragard **(25)** studied welded joints of steel with lower yield strengths for $R = 0$ and -1. The results of these two authors show that the improvement due to TIG dressing is not affected by the R ratio (Fig. 9).

Thus, for fillet welds for high strength structural steels (yield strength >500 MPa), fatigue improvement due to TIG dressing is approximately 100 percent over a wide range of R values (from -3 to 0.4). For very high R ratios e.g. 0.7, a smaller improvement is observed. For steels with a yield strength of 400 MPa, the improvement is 70 percent for $R = -1$ and 0; for steels with a yield strength of 250 MPa, the improvement is 40 per cent for $R = -1$ and 0.

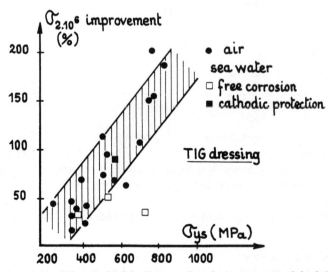

Fig 8   Influence of the yield strength of the base metal on the improvement of the fatigue strength by TIG dressing (3)

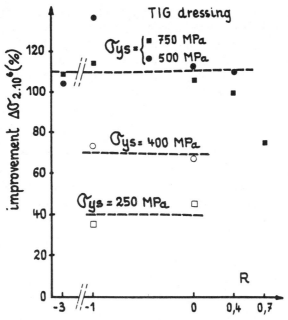

Fig 9 Influence of the *R* ratio on the improvement of fatigue strength of TIG dressed joints for different steel grades (3)

Effect of seawater

Currently, few results are available with regard to the effectiveness of TIG dressing technique in the presence of seawater. German results **(12)** under variable amplitude loading at 10 Hz show good corrosion fatigue strength for

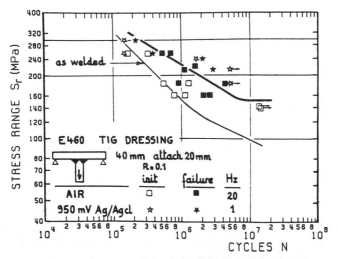

Fig 10 Fatigue strength for fillet welds with TIG dressing (4)

welded joints of high strength structural steels (E355, E460, E690) subjected to TIG dressing, under free corrosion conditions. French results (4) for E460 steel welded joints in seawater with cathodic protection (-950 mV/Ag/AgCl) reveal lifetimes at 1 Hz comparable to those obtained in air. The fatigue strength improvement over 2×10^6 cycles is 45–50 percent in relation to as-welded improved profile assemblies.

Inspection

In order to ensure that TIG dressing is correctly performed, qualified personnel must be trained for this type of work, as is the case for welding. Initial tests should also be performed in order to determine optimal operating conditions. Since TIG dressing appearance is usually very clean, visual inspection easily enables defects to be detected. Geometric appearance is of prime importance.

Weld toe grinding

Operating procedure

Grinding is usually introduced to eliminate defects and to introduce a longer crack initiation period by removing metal from the weld toe. This machining can be performed using a disk grinder but preferably with a rotary burr grinder. To eliminate all slag inclusions and microcracks, machining depth should be 0.5–0.8 mm (19). While eliminating crack initiation sites, this technique enables the weld toe profile to be improved, thus reducing the local stress concentration coefficient.

Results

Available results show that fatigue strength is improved when local machining is performed. The improvement obtained over 2×10^6 cycles varies from 30 to 100 percent. Fatigue strength improvement rates for ground joints are shown in Fig. 12, in relation to base metal yield strength.

Air testing

In Dutch work (2), 40 mm thick 'T' joints with ground weld toes were subjected to fatigue loading. The results show a clear improvement (50 percent) in relation to studies involving as-welded joints. Similar work in the UK (6) involved 38 and 25 mm thick cruciform joints. Milling with a disk grinder (100 mm diameter 36 grit in an epoxy matrix) was used to improve the welds. The results are comparable to those of the Dutch programme. The fatigue strength improvement over 2×10^6 cycles is, respectively, 70 percent and 110 percent for the 38 and 25 mm thicknesses, Fig. 11.

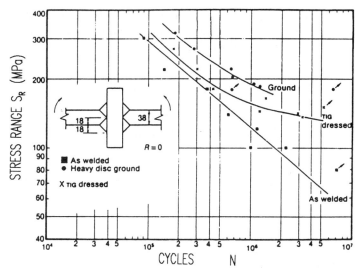

Fig 11 Fatigue strength for ground fillet welds (6)(8)

Seawater testing

Seawater testing at a frequency of 0.2 Hz was performed. The results, although less encouraging than those for air testing, showed marked fatigue strength improvement due to weld toe grinding. Over 2×10^6 cycles, the seawater fatigue strength was improved by 20–30 percent for 40 mm thick joints. Figure 12 gives the results of the British programme (6). Additional work by Booth (8) studied the effect of cathodic protection on ground joints. The results

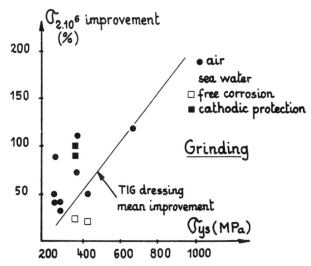

Fig 12 Influence of the base metal yield strength on the improvement of fatigue strength by grinding (3)

obtained suggest that cathodic protection enables the same fatigue strength characteristics as for air to be obtained, even in the case of assemblies that had been corroded prior to testing (see Fig. 12).

Inspection

Grinding inspection should be based on the appearance of the area that has been ground. The geometry obtained should result in better stress distribution at the weld toe. The DoE Guidance Notes (18) recommend that the depth of grinding should be 0.5 mm below the bottom of any visible undercut and should not exceed 2 mm or 5 percent of the plate thickness (Fig. 13). No score marks likely to be sites of fatigue crack initiation should exist. Inspection can be completed by magnetic particle control.

Surface prestressing

Weld toe hammer peening

This technique differs greatly from the others in that a large residual compressive stress is introduced to the surface. Hammer peening modifies the geometry of the weld bead. After hammer peening, defects are surrounded by a volume of cold-worked material with a high residual compressive stress.

Operating procedure, inspection

This improvement technique consists of cold hammering the weld toe using a pneumatic hammer. The hammer may be equipped with a hemispherical head tool or a needle device. Hammer peening effectiveness depends upon the number of passes and the duration of the operation. If hammer peening is too rapid, the depth of the deformed area may be insufficient to surround all defects with a field of residual compressive stress. Generally speaking, an indentation depth of 0.6 mm is a good compromise between treatment time and effectiveness (19).

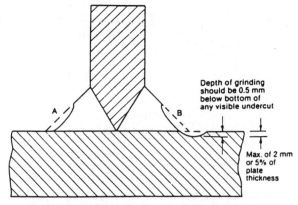

Depth of grinding should be 0.5 mm below bottom of any visible undercut

Max. of 2 mm or 5% of plate thickness

Fig 13   Recommendation for toe grinding by the DoE (10)

Results

The results obtained using this method are remarkable (Fig. 14). In relation to TIG dressing, improvement is twice as great. Here again, the higher the yield strength, the greater the improvement (Fig. 15). The effect of the R ratio was studied by Booth (7). His results for 12.5 mm cruciform joints show that the improvement is 170 percent for $R = -1$, and 80 percent and 30 percent, respectively, for $R = 0$ and 0.5 when low yield strength steels are used. The

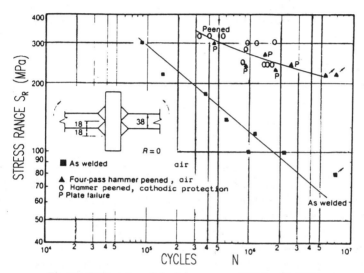

Fig 14   Fatigue strength for hammer-peened fillet welds (6)(8)

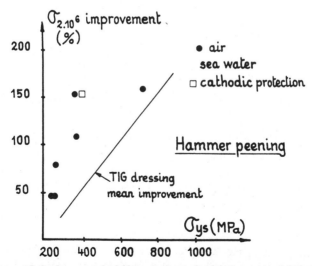

Fig 15   Influence of the base metal yield strength on the improvement of fatigue strength by hammer peening (3)

British programme (6) resulted in a fatigue strength improvement of 100–150 percent for 25 and 38 mm thick cruciform joints. Booth's results (8) show that in seawater if cathodic protection is used, fatigue strength of hammer peened joints is comparable to that obtained in air, Fig. 14.

Shot peening

The surface to be treated is sprayed with steel shot. Each impact has the same effect as hitting the surface with a small hammer. This treatment enables a residual compressive stress field of approximately 0.75 the yield strength to be introduced into the surface layer.

Operating procedure

The purpose of shot peening is to create compressive pre-stressing in the surface layer; the stress level, uniformity, and the depth of the pre-stressed layer must be reproducible.

This operation induces surface residual compressive stress, especially in stress concentration areas (e.g., the weld toe). Surface defects must be included in this residual stress field. Conditions that induce a maximum depth affected by the shot peening should be obtained. This is the case for high ALMEN intensities, obtained using large shot. As pointed out by Bignonnet et al. (5) the increase in ALMEN intensity results primarily in larger affected depth, but the level of residual stress remains roughly the same; Hoffman and Muesgen (16) show that optimum ALMEN intensity depends on steel grade. The higher the yield strength the higher the optimum ALMEN intensity. Other factors are also important: all surface defects (undercuts, etc.) must be eliminated by the peening conditions used. This can be achieved using small shot and varying the angle of incidence in relation to the weld. Two main requirements affect the shot peening of welded joints:

– maximum affected depth using a large ALMEN intensity (Fig. 16).
– treatment of all defects using small shot.

Two possible methods of achieving an acceptable compromise have been proposed (5).

(a) Single-stage shot peening requires careful selection of shot that is as large as possible yet compatible with the size of the defects to be treated. Statistical distribution of defect size on the weld surface must be determined. For welds with good geometry, S330 shot can be used; it induces a fairly high ALMEN intensity. For poorer quality welds (undercuts, sharp weld bead, etc.), smaller shot must be used (S230, S170, etc.). Low intensities are induced, and peening effectiveness may be reduced.

(b) Two-stage peening may also be performed. An initial peening using small shot (or appropriate sandblasting) is used for treatment and to eliminate sharp, deep defects. A second pass using larger shot is performed to

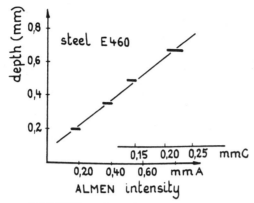

Fig 16 Influence of ALMEN intensity on the affected depth by residual stresses (5)

achieve a high ALMEN intensity, i.e., reach a greater affected depth. The aim of double peening is to reduce the scatter in fatigue and so to reach an increased reliability and confidence in shot peening of welded structures. Examples of shot-peening parameters used for fatigue tests are given in Table 2.

Table 2   Examples of shot peening conditions

Reference		(5)	(4)		(12)	(8)	
Steel		E 460/550	E 460		E 355/460/620	E 355	
Nb run		1	2		1	1	
			First	Second			
Shot		S 330	S 110	S 550	S 230	S 230	S 330
Mean ϕ	mm	0.85	0.3	1.4	0.55	0.55	0.85
Almen int.	mm	0.50, 0.55A	0.15, 0.20A/0.20, 0.23C		0.27A	0.4, 0.5A	0.6, 0.7A
Coverage	percent	200	125	125			

Results

Maddox (**22**) studied the effect of shot penning on two normalized grades of steel and two quenched and tempered steel grades. Two types of test specimens were used: cruciform joints and joints with longitudinal reinforcement. For normalized steel longitudinal joints (yield strength = 260 and 390 MPa), improvement is approximately 35 percent; it is approximately 70 percent for quenched and tempered steel (yield strength = 730 and 820 MPa). For cruciform joints, the improvement is much greater: approximately 90 percent for normalized steels. Lieurade and Bignonnet (**21**)(**15**) also demonstrated an improvement of 60 percent for butt welds of E355 and E490 steel, and 40 percent for 'T' joints of 30 mm thick E460 steel (Figs 17 and 18). The German program (**12**) provides results under variable amplitude loading for shot peened joints subjected to seawater and free corrosion. The results for three steels (E355, E460, E690) show no harmful effects due to variable amplitude loading, which indicates that shot peening stress relieving is not a problem,

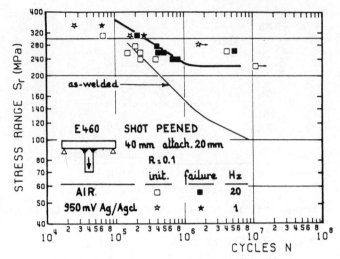

Fig 17   Fatigue strength for shot peened fillet welds (4)

even under these conditions. Given these results shot peening does not seem to be affected by seawater **(14)(12)**.

Inspection

Shot peening is still a somewhat empirical method. Complex phenomena affecting treated material behaviour make it difficult to determine optimal technical parameters. These parameters are numerous, and sometimes inter-dependent, for example: shot (glass, steel, ceramic, hardness, dimension);

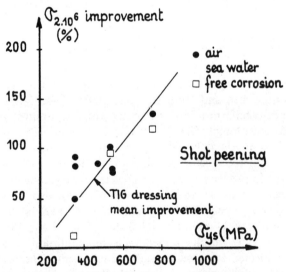

Fig 18   Influence of the base metal yield strength on the fatigue strength improvement by shot peening (4)

machine (nozzle or turbine position, nozzle type, air pressure, turbine speed); peening duration – coverage rate.

In order to deal with the multiple parameters, a method of controlling shot peening repetitiveness was developed; an ALMEN strip (11) was used. The deflection of the standard strip under shot projection defines the ALMEN intensity. At a first approximation the ALMEN intensity is proportional to the cold working depth. To obtain the required ALMEN intensity, preliminary shot peening tests are performed using an ALMEN strip-holder frame. The frame represents the area to be treated. Several ALMEN strips are arranged on the frame so that they are in the same position as the welds to be treated. Optimal equipment conditions (nozzle diameter, air pressure, nozzle part distance, etc) are set so as to obtain the required ALMEN intensity.

The second controlled shot peening parameter is the coverage rate. This is done using a fluorescent dye sprayed on before shot peening. During peening, the shot removes the dye; after peening, it is easy to determine whether the surface has been treated completely.

To obtain effective shot peening, the entire critical surface must be treated. A coverage rate of 100 percent is achieved for a peening time such that the entire surface is uniformly covered by shot impact. When this time is doubled, a coverage rate of 200 percentage is obtained.

A coverage rate of 200 percent is usually used since it is a good compromise between treatment time and effectiveness. After treatment, magnetic particle control of the treated welds may be performed.

Conclusions

The main purpose of each and every improvement treatment is to safeguard against fatigue-induced cracking in welded joints; in other words, to introduce a crack initiation phase representing a major portion of the structure's total lifetime. The only way to obtain this result is to eliminate weld defects and considerably reduce effective local stress concentrations by remodelling weld geometry (improved profile, grinding, TIG dressing), and/or inducing a high residual compressive stress field into the sensitive area (hammer and shot peening). Use of a technique inducing residual compressive stress after the geometry has been improved may result in a marked increase in fatigue strength.

The following conclusions can be drawn from test results.

(1) All techniques discussed in this paper produced encouraging results, with fatigue strength improvement ranging from 20 to more than 100 percent in air.

(2) In seawater, under free corrosion conditions, the improvement rate was lower than in air (early crack initiation by corrosion pitting); however, fatigue strength remained slightly greater than that of the as-welded joint.

(3) Using cathodic protection enabled the benefits of the improvement treatment to be maintained, and lifetimes were the same as in air; in low stress fields, longer lifetimes than in air were regularly observed.

(4) Testing under variable amplitude loading confirmed the benefits of weld improvement techniques, including methods that induce residual compressive stress.

(5) Improving the weld geometry reduces the thickness effect substantially.

(6) Generally speaking, improvement increased with yield strength.

(7) The success and ease of postweld treatments increased with the care taken in making the as-welded assembly. It should also be remembered that the weld toes is not the only site for crack initiation; at the same time as the weld toe is being improved one should verify that other causes of rupture (lack of penetration, internal defects, too-short weld beads, etc.) have been eliminated so as to obtain good fatigue strength.

Use of the above techniques is subject to correct control. Improvement is directly linked to the care taken when applying the treatment; this work should be performed by properly trained and qualified personnel. The transition in the weld toe area should be as gradual as possible; any defects likely to cause crack initiation must be eliminated or made harmless. In all cases, preliminary tests must be performed in order to determine the operating conditions that will result in optimal weld improvement.

Finally, using high strength steels with appropriate weld finishing techniques in the sensitive areas allows the engineer to use the advantages of such materials, while benefiting from their excellent fatigue strength compared to conventional steels.

References

(1) AWS D1 1. (1980) *Structural Welding Code*, American Welding Society.

(2) DE BACK, J., VAESSEN, G. H. G. *et al.* (1981) Fatigue and corrosion fatigue behaviour of offshore steel structures, ECSC Convention 7210 KB/602, Final report.

(3) BIGNONNET, A. (1987) Improving the fatigue strength of welded steel structures, *Steel in marine structures*, (Edited by C. Noordhoek and J. de Back); Elsevier Science, pp. 99–118.

(4) BIGNONNET, A., PAPADOPOULOS, Y., BARRÈRE, F., LIEURADE, H. P., and LECOQ, H. (1987) The influence of cathodic protection and post weld improvement on the fatigue resistance of steel welded joints, *Steel in marine structures*, (Edited by C. Noordhoek and J. de Back); Elsevier Science, pp. 737–746.

(5) BIGNONNET, A. PICOUET, L., LIEURADE, H. P. and CASTEX, L. (1987) The application of shot peening to improve the fatigue life of welded steel structures, *Steel in marine structures*, (Edited by C. Noordhoek and J. de Back); Elsevier Science, pp. 669–678.

(6) BOOTH, G. S. (1978) Constant amplitude fatigue tests on welded steel joints performed in air, *European offshore steel research seminar*. The Welding Institute, Cambridge.

(7) BOOTH, G. S. (1981) The fatigue life of ground and peened fillet welded steel joints. The effect of mean stress, *Met. Construction*, **13**, 112–115.

(8) BOOTH, G. S. (1987) Techniques for improving the corrosion fatigue strength of plate welded joints, *Steel in marine structures*, (Edited by C. Noordhoek and J. de Back); Elsevier, pp. 747–758.

(9) DIJKSTRA, O. and NOORDHOEK, C. (1985) The effect of grinding and special weld profile on the fatigue behaviour of large scale tubular joints, Proceedings of the Offshore Technology Conference.

(10) Department of Energy (1984) *Offshore installations. Guidance on Design and Construction*, HMSO, London, UK.

(11) FUCHS, H. O. (1984) Defects and virtues of the ALMEN intensity scale, Second International Conference on Shot Peening, (Edited by H. P. Fuchs). The American Shot Peening Society, pp. 74–78.

(12) GRIMME, D., *et al.* (1984) Untersuchungen zur Betriebsfestigkeit von geschweissten offshore – konstruktionen in Seawater, ECSC Agreement 7210 KG/101.

(13) HAAGENSEN, P. J. (1979) TIG dressing of steel weldements for improved fatigue performance, Proceedings of the Offshore Technology Conference.

(14) HAAGENSEN, P. J. (1981) Improving the fatigue of welded joints. Proceedings of the International Conference on *Steel in Marine Structures*, pp. 381–442, IRSID, France.

(15) HAAGENSEN, P. J., DRAGEN, A., SLIND, T., and ORJASOETER, O. (1987) Prediction of the improvement in fatigue life of welded joints due to grinding, TIG dressing, weld shape control and shot peening, *Steel in Marine Structures* (Edited by C. Noordhoek and J. de Back); Elsevier, pp. 689–698.

(16) HOOFMAN, K. and MUESGEN, B. (1987) Improvement of the fatigue behaviour of welded high strength steels by optimized shot peening, *Steel in Marine Structures*, (Edited by C. Noordhoek and J. de Back); Elsevier, 679–688.

(17) IIDA, K. and ISHIGURO, T. (1977) Brief summary of Japanese documents concerned with improvement of fatigue strength of welded joints, International Institute of Welding, Doc IIW–XIII–862–77.

(18) KANAZAWA, S., ISHIGURO, T., HANZAWA, M., and YOKOTA, H. (1979) The improvement of fatigue strength in welded high tensile strength steels, International Institute of Welding IIW–Doc. XIII–735–74.

(19) KNIGHT, J. W. (1977) Improving the fatigue strength of fillet welded joints by grinding and peening, International Institute of Welding, Doc IIW–XIII–851–77.

(20) KOBAYASHI, K. *et al.* (1977) Improvement in the fatigue strength of fillet welded joint by use of the new welding electrode, International Institute of Welding Doc. IIW–XIII–828–77.

(21) LIEURADE, H. P. and TOURADE, J. C. (1983) Effect des traitements de parachèvement sur le comportment de joints soudés bout à bout en acier HLE, *Mém. et Et. Sci. Revue de Métallurgie*, **9**, 455–467.

(22) MADDOX, S. J. (1982) "Improving the fatigue lives of fillet welds by shot peening", Proceedings of the IABSE Colloqium, Lausanne, pp. 377–385.

(23) MILLINGTON, D. (1973) TIG dressing for the improvement of fatigue properties in welded high strength steels, International Institute of Welding IIW–DOC. XIII–698–73.

(24) MINNER, H. H. and SEEGER, T. (1979) Investigation of the fatigue strength of weldable high strength steels St 460 and St E490 in as-welded and TIG dressed conditions, International Institute of Welding–Doc. IIW–XIII–912–79.

(25) SIMON, P. and BRAGARD, (1978) Amélioration des propriétés de fatigue des joints soudés, Convention n° 6210-45/2/202. Commission des Communautés Européennes.

Y. Mutoh* and M. Takeuchi†

The Effect of the Coating Layer Thickness on High Temperature Fatigue Strength in Thermal Barrier-Coated Steel

REFERENCE Mutoh, Y. and Takeuchi, M., **The effect of coating layer thickness on high temperature fatigue strength in thermal barrier-coated steel,** *Fatigue Design*, ESIS 16 (Edited by J. Solin, G. Marquis, A. Siljander, and S. Sipilä) 1993, Mechanical Engineering Publications, London, pp. 157–169.

ABSTRACT The fatigue strengths of ceramic-coated steel at room temperature and 773 K were almost identical. The fatigue strength of the specimen with a thick ceramic-coating layer was lower than that with a thin ceramic-coating layer. These results are conceivable, considering that the thickness of the ceramic layer corresponds to an initial defect size.

Introduction

Ceramic coatings effectively utilize the excellent properties of ceramics, such as high temperature resistance, excellent corrosion resistance, and wear resistance etc., while avoiding limitations such as poor toughness and high manufacturing costs. The properties required for ceramic-coating materials are: low thermal conductivity, high temperature stability, high heat radiancy, and a similar thermal expansion coefficient to the substrate.

The high temperature-resistance properties of ceramic coatings were evaluated in burner rig tests (1)–(4) and thermal cycle tests (5)–(8). However, in these tests no external stresses are applied to the test specimens since ceramic coatings are currently only applied to those components which are exposed to low external stresses. To extend their application to a wide range of components used in high temperature and high applied stress environments, the high temperature fatigue characteristics of ceramic coatings need to be investigated (9)(10). Moreover, in the burner rig and thermal cycle tests the only material employed as a substrate was a nickel-based alloy, which was used mainly for a gas turbine. Although there is an increasing demand for the application of ceramic coatings for reactor vessels and high efficient boilers in which structural steels are mainly used, sufficient attention has not been paid to the high temperature resistance characteristics or fatigue behaviour of ceramic-coated structural steels.

In this study, the fatigue performance of ZrO_2 plasma-spray coated structural steel at room and elevated temperatures was assessed. The fatigue fracture processes and the effect of the coating layer thickness on fatigue behaviour were investigated.

* Nagaoka University of Technology, Nagaoka-shi 940-21, Japan.
† Honda UK Manufacturing Limited, Swindon, Wiltshire, SN3 4TJ, UK.

Experimental procedures

Specimens

The material used as a substrate was a structural carbon steel (JIS S45C). Its chemical compositions and mechanical properties are shown in Tables 1 and 2, respectively. The dimensions of the fatigue specimen are shown in Fig. 1. Three different specimens, namely an uncoated specimen and two ceramic-coated specimens with a ceramic-coated layer thickness of 0.1 mm and 0.3 mm were tested. The gauge part of the uncoated specimen was polished with 600 grit abrasive silicon paper and cleaned with acetone prior to the tests. The ceramic-coated specimens were made as follows; the gauge part of the specimen was shot-blasted with alumina grit prior to the plasma spraying. A NiCrAlY alloy, which is the material suitable for a bonding layer between the ceramic coating and the metallic substrate, was plasma-sprayed uniformly onto the blasted area, and then zirconia was plasma-sprayed onto the NiCrAlY alloy bond-coating layer. The spraying conditions of the NiCrAlY alloy and the zirconia are shown in Table 3. The thickness of the NiCrAlY alloy layer was 0.1 mm. The ceramic layers were 0.1 mm and 0.3 mm in thickness. Figure 2 shows the longitudinal cross-sections of the coating layers.

Fatigue test

The specimens were tested at room temperature and 773 K using a servo-hydraulic type fatigue machine with a stress ratio of -1 and a frequency of

Table 1   Chemical compositions of JIS S45C steel used (of wt%)

C	Si	Mn	P	S	Ni	Cr	Mo	Cu
0.41	0.22	0.68	0.022	0.018	<0.01	0.02	<0.01	0.01

Table 2   Mechanical properties of JIS S45C steel used

Temp. (K)	Yield strength (MPa)	Tensile strength (MPa)	Elongation (%)	Reduction of area (%)	Vickers hardness
RT	461	718	26.8	54.5	190
773	225	379	22.5	83.5	—

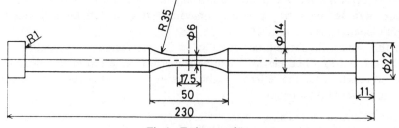

Fig 1   Fatigue specimen

Table 3 Plasma spraying conditions

	Bond coating	Ceramic coating
Powder	Ni–17.1 wt% Cr–5.5 wt% Al–0.65 wt% Y	ZrO_2–7 wt% CaO
Grain size (μm)	44 ~ 10	44 ~ 10
Arc current (A)	750	850 $\pm$ 30
Arc voltage (V)	35	32 ~ 33
Arc gas	Ar/He	Ar/He
Gas pressure (MPa)	0.41/0.34	0.27/0.34
Spraying distance (mm)	120	90
Revolutions of test piece (rpm)	750	750

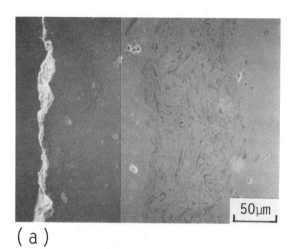

(a)

(b)

Fig 2 Observations of longitudinal cross-sections of the coated layers. (a) $t = 0.1$ mm; (b) $t = 0.3$ mm

5 Hz. An electric furnace was used for the high temperature fatigue tests. The test temperature was measured using an alumel–chromel thermo-couple attached as near as possible to the gauge part of the specimen. The distribution of temperature along the gauge part of the specimen was in the range 771 K–775 K, and the fluctuation of the temperature during the test was ± 1 K. The test was started 1 h after the temperature reached 773 K. The fracture surfaces were examined in a scanning electron microscope in detail.

FEM analysis

The axi-symmetric thermo-elastic–plastic stress analysis of the ceramic-coated specimens under tensile loading and during heating was carried out using the general FEM program MARC, where residual stress induced in the coating layer during the coating process was not considered.

Double the length (12 mm) of the diameter was considered in an axial direction in order to calculate the longitudinal stress distribution around the centre of the gauge length. The other dimensions of the modelled specimen for FEM analysis, such as diameter and coating layer thicknesses, were identical to the coated specimen used in the experiments. The FEM analysis was made on a quarter of the coated specimen due to the symmetry of the specimen. As shown in Fig. 3, the overall area of the finite element model was divided into eight blocks. In the z (axial) direction, blocks 1–4 were divided into eleven equal parts and blocks 5–8 were divided into five equal parts. In the r direction, blocks 1 and 5 were bisected and blocks 3, 4, 7, and 8 were divided into

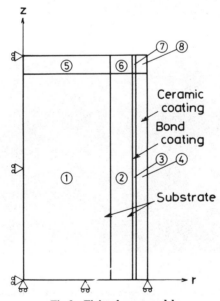

Fig 3   Finite element model

Table 4 Physical properties of the materials used in the FEM analysis

Material	Temp. (K)	Young's modulus (GPa)	Poisson's ratio	Yield strength (MPa)	Thermal expansion coefficient ($\times 10^{-6}/K$)
Ceramic	RT	11.0	0.25	—	5.4
coating	773	11.0	0.25	—	5.4
Bond	RT	200.0	0.31	597	12.1
coating	773	152.5	0.31	427	13.2
S45C	RT	210.0	0.28	461	11.2
substrate	773	170.0	0.28	225	14.0

three equal parts. The resultant numbers of elements and nodes were 144 and 504, respectively. In the boundary conditions the nodes on the z axis were fixed in the r direction and the nodes on the r axis were fixed in the z direction.

The physical properties of the materials employed in the calculation are listed in Table 4, where those of the ceramic coating and the bond coating were obtained from references (10) and (11). It was assumed that the Young's modulus, yield strength, and thermal expansion coefficient had a linear relation with temperature in the range of room temperature to 773 K. The work hardening coefficient of S45C was assumed to be unaffected by temperature and 73.4 GPa was employed based on the experimental results. Since the work hardening coefficient of the bond coating material was still unknown, the bond coating material as well as the ceramic coating material was assumed to be the elastic body.

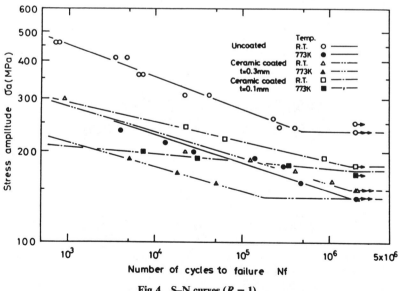

Fig 4 S–N curves ($R = 1$)

Table 5   Fatigue limits

Materials	Temp. (K)	Fatigue limit (MPa)	Ratio of fatigue limits*
S45C	RT	230	—
substrate	773	140	—
Coated	RT	150	0.65
(t = 0.3 mm)	773	140	1.00
Coated	RT	180	0.78
(t = 0.1 mm)	773	170	1.21

* Ratio of fatigue limits of S45C substrate and coated specimens.

Results

Fatigue life

The S–N curves determined at room temperature and 773 K are shown in Fig. 4. The fatigue limits determined from the curves are listed in Table 5. At room temperature the fatigue life and limit were reduced by ceramic coating. At high temperature the fatigue life at the higher stress level was longer for the uncoated specimen than for the ceramic-coated specimens. At the lower stress level both the fatigue life and limit of the ceramic-coated specimen were almost the same as or higher than those of the uncoated specimens. It can also be noted that on the uncoated specimen the fatigue limit was significantly reduced at high temperature, but on the ceramic-coated specimens the fatigue limit was hardly affected by temperature. The effect of coating thickness on fatigue life was less remarkable at the higher stress level, but at the lower stress level the thicker the coating, the lower the fatigue life and limit.

Fractographic observations

Figure 5 shows the macroscopic view of the ceramic-coated specimens tested at room and high temperatures. As can be seen, on the coated specimen there

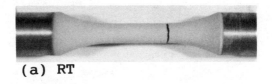

(a) RT

(b) 773 K

Fig 5   Macroscopic view of fractured specimens

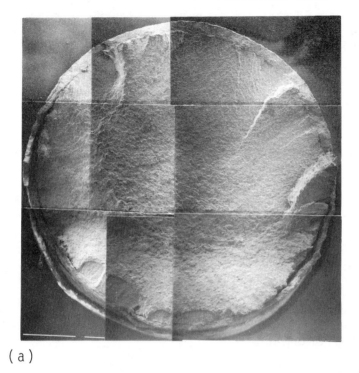

(a)

(b)

Fig 6   Fractographs of the coated specimen tested at 773 K. (a) General view of fracture surface. (b) Higher magnification of crack initiation region

was no visible difference between room and high temperatures, but the uncoated specimen was significantly oxidized at high temperature. The fracture surfaces of both the specimens were oxidized and, moreover significant oxidation was observed in the fatigue fracture initiation region of the uncoated specimen.

Figure 6 shows the fracture surface of the ceramic-coated specimen tested at 773 K. A separation of the ceramic-coating layer from the substrate was observed on most of the area except on that where the crack was initiated. It seems that the separation occurred at the final unstable fracture. The fracture surface shown in Fig. 6(b), indicates that continuous crack growth from the coating layer into the substrate occurred. Similar results were observed at all stress levels regardless of temperature and ceramic layer thickness.

Results of the FEM analysis

Figure 7 shows the radial distribution of the z direction thermal stress induced by isothermal heating from room temperature to 773 K. It can be seen that the increase of temperature from room temperature to 773 K produces a tensile stress of 40–50 MPa in the ceramic layer. The result shown in Fig. 7 indicates that the thermal stress may contribute to the reduction in fatigue strength of the ceramic-coated specimen. According to this hypothesis the thickness of the coating does not have a significant effect on the thermal stress. Figure 8 shows the radial distribution of the z direction stress induced when the tensile elongation of 5.6×10^{-6} mm is applied to the coated specimen. In analysis neither an initial residual stress nor stress relaxation in the substrate during tests were considered. The stress induced in the ceramic layer due to the difference of the Young's modulus between the coating-layer and the substrate was found to be approximately one-twentieth of the stress induced in the substrate. There was no effect on coating thickness.

Discussion

Fatigue fracture process in the coated specimen

In the fatigue fracture process of an uncoated specimen, microcracks are initi-

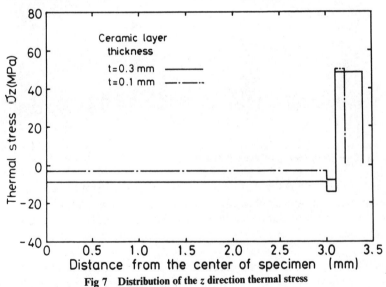

Fig 7   Distribution of the z direction thermal stress

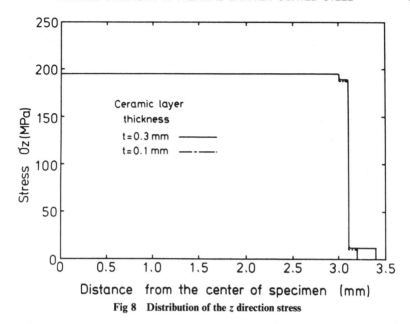

Fig 8   Distribution of the z direction stress

ated at localized slip bands formed on the metal surface and grow to fatigue cracks (12). However, on a ceramic-coated specimen it is thought that different fatigue processes occur since neither plastic deformation of the coating layer nor slip bands on the coating surface occur. From the observations of the longitudinal cross-sections of the coated specimens fatigued as shown in Fig. 9, the fatigue cracks were propagated from the coated surface. From the fractographic observations mentioned above, in the fatigue fracture initiation

Fig 9   Fatigue crack in the ceramic layer

region the fracture surface at the interface between the coating layer and the substrate was smooth and no initiation of internal cracks independent of the cracks from the ceramic coating layer was found. These results indicate that the initiation of the fatigue crack in the coated specimen occurs on the surface of the ceramic coating and the crack propagates into the substrate.

Although an accurate observation of cracks on the coating surface was not possible due to the roughness of the sprayed surface, it is thought that the fatigue cracks are initiated in the early stage, in some cases, at less than 2 percent of the total fatigue life, as shown in Fig. 10. This early crack initiation is extremely different from the case of the uncoated plain specimen. The rough as-sprayed surface and the porous ceramic coating layer with microcracks would contribute to the early stage of crack initiation.

The effect of ceramic layer thickness

As shown in Fig. 4, the fatigue strength of the thinner coated specimen is higher than that of the thicker coated specimen at both room temperature and 773 K. Although there are many factors affecting the fatigue strength of the ceramic-coated specimen, the factors such as (1) residual stress, (2) distribution of the loaded stress, (3) materials of the coating layer, and (4) roughness of the coating surface are thought to be dominant.

The residual stress induced in the layer may be reduced due to the formation of multiple microcracks during the spraying and subsequent cooling down processes **(13)**. Based on this and the results shown in Fig. 7, no significant changes in residual stress occur in the range of 0.1–0.3 mm ceramic layer thickness, which is relatively small considering the diameter of the specimen. The distribution of the applied stress is not affected by the ceramic layer thickness in the present test conditions, as shown in Fig. 8. The coating layer materials for the specimens with a ceramic layer thicknesses of 0.1 and 0.3 mm can be thought to be identical, namely, the hardness of the former is 677 and that of the latter 683 Hv and also both have a similar porosity level. The surface roughness of the two coatings is similar, namely R_{max} 34–35 μm.

As discussed above, the effect of coating thickness on the fatigue strength cannot be explained by the factors (1)–(4). According to the fatigue fracture process proposed above, the crack is initiated in the ceramic layer in the early

Fig 10   Fatigue crack observed at the early stage of fatigue life (2% N_f)

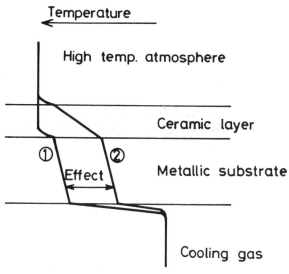

Line ①: without ceramic coating

Line ②: with ceramic coating

Fig 11 Thermal barrier effect of ceramic coating

stage and subsequently propagates into the substrate. Furthermore, compared with the crack growth curves for metallic materials, those for ceramic materials lie in the much higher region of crack growth rate and their slope is greater. The exponent m in the crack growth curve $da/dN = C(\Delta K)^m$ is in the range of 2–4 for the metal and 10–30 for the ceramics, i.e., one order greater for ceramics (14)(15). These facts indicate that once the crack initiates, it grows rapidly in the ceramic layer. The observation of the longitudinal cross-section of the coated specimen, which was tested at the fatigue limit, confirmed the existence of the crack in the ceramic layer. Therefore, as the first stage of approximation, it is assumed that the initial crack which already exists in the ceramic layer and the depth of which is equal to the ceramic layer thickness, propagates into the substrate. The difference in the ceramic layer thickness between the two specimens results in the difference in the initial defect size. Although the stress intensity factor at the interface between the two different materials is not known, it is thought to be qualitatively proportional to the square root of the crack length. Assuming that the initial crack is circumferential and the threshold value of the stress intensity factor for the crack propagation from the interface is independent of the ceramic layer thickness, in the case of the 0.1 mm ceramic layer thickness, a 1.7 times higher stress than the case of the 0.3 mm ceramic layer thickness is required as shown in the following equation. The factors of 1.18 and 1.14 in the equation correspond to the boundary correction factors for circumferential cracks (16).

$$\sigma_{0.1} = (1.18\sqrt{\pi} \cdot 0.3/1.14\sqrt{\pi} \cdot 0.1)\sigma_{0.3}$$

$$\fallingdotseq 1.8\sigma_{0.3}$$

However, the actual fatigue strength of the specimen with 0.1 mm ceramic layer thickness was 1.2 times higher than that of the specimen with a 0.3 mm ceramic layer thickness. The assumption of a circumferential initial crack may be too severe. This also means that the factors such as crack growth in the ceramic layer, variation of the K value near the interface between the ceramic layer and the substrate, and the initial residual stress need to be considered.

The thermal barrier effect of the ceramic coating is an increasing thickness of the ceramic layer, as schematically illustrated in Fig. 11. The fatigue strength, however, is reduced as the thickness of the ceramic coating increases. Therefore, the thickness of the ceramic layer should be determined in order to clarify these two contradicted requirements.

Conclusion

The following results can be summarised from the present work.

(1) The fatigue strength of ceramic-coated steel was almost identical at both room temperature and 773 K. It was lower than that of uncoated steel at room temperature, but it was higher at 773 K.

(2) The fatigue fracture processes of the ceramic-coated steel were as follows: (a) initiation of a fatigue crack on the surface of the coating in the very early stage of fatigue; (b) rapid propagation of the fatigue crack in the coating layer; (c) penetration of the fatigue crack into the substrate. There was no further initiation of other fatigue cracks at the interface between the coating layer and the substrate.

(3) The fatigue strength of the specimen with a thicker coating layer was lower than that with a thinner coating layer. This result is true if the thickness of the ceramic layer corresponds to an initial defect size.

References

(1) McDONALD, G. and HENDRICKS, R. C. (1980) Effect of thermal cycling on $Z_1-O_2-Y_2O_3$ thermal barrier coatings, *Thin Solid Films*, **73**, 491.

(2) MILLER, R. A. and LOWELL, C. E. (1982) Failure mechanisms of thermal barrier coatings exposed to elevated temperatures, *Thin Solid Films*, **95**, 265.

(3) WATSON, J. W. and LEVINE, S. R. (1984) Deposition stress effects on the life of thermal barrier coatings on burner rigs, *Thin Solid Films*, **119**, 185.

(4) JOHNER, G. and SCHWEITZER, K. K. (1985) Flame rig testing of thermal barrier coatings and correlation with engine results, *J. Vac. Sci. Technol*, **3A-6**, 2516.

(5) STECURA, S. (1982) Two-layer thermal barrier systems for Ni-Al-MO alloy and effects of alloy thermal expansion on system life, *Am. Ceram. Soc. Bull.*, **61**, 256.

(6) SHANKER, N. R., *et al.* (1983) Acoustic emission from thermally cycled plasma-sprayed oxides, *Am. Ceram. Soc. Bull.*, **62**, 614.

(7) STECURA, S. (1980) Effects of yttrium, aluminium, and chromium concentrations in bond coatings on the performance of zirconia–yttria thermal barriers, *Thin Solid Films*, **73**, 481.

(8) BERNDT, C. C. and HERMAN, H. (1983) Failure during thermal cycling of plasma-sprayed thermal barrier coatings, *Thin Solid Films*, **108**, 427.

(9) SCHNEIDER, K. and GRUNLING, H. W. (1983) Mechanical aspects of high temperature coatings, *Thin Solid Films*, **107**, 395.

(10) KAUFMAN, A., LIEBERT, C. H., and NACHTINGALL, A. J. (1978) Low-cycle fatigue of thermal-barrier coatings at 982°C, NASA TP. 1322.

(11) SIEMERS, P. A. and HILLIG, W. B. (1981) Thermal barrier coated blade study, NASA CR. 165351.

(12) FROST, N. E., MARSH, K. J., and POOK, L. P. (1974) *Metal fatigue*, Chapter 2, Oxford University Press.

(13) NOUTOMI, A., *et al.* (1988) Study on residual stress measurement of plasma sprayed coatings, *Q. J. Jpn Welding Soc.*, **6**, 341.

(14) DAUSKARDT, R. H. and RITCHIE, R. O. (1991) Cyclic fatigue-crack propagation in ceramics and ceramic composites, *Mechanical behaviour of materials-VI*, Vol. 2, Pergamon Press, Oxford. p. 325.

(15) MUTOH, Y. *et al.* (1991) Fatigue crack growth of long and short cracks in silicon nitride, *Fatigue of advanced materials*, MCEP, Birmingham, p. 211.

(16) *Stress intensity factors handbook* (1987) (Edited by Y. Murakami, *et al.*), Pergamon Press, Oxford, p. 644.

G. Härkegård* and S. Stubstad†

Simplified Analysis of Elastic–Plastic Strain Concentration in Notched Components under Cyclic Loading

REFERENCE Härkegård, G. and Stubstad, S., **Simplified analysis of elastic–plastic strain concentration in notched components under cyclic loading,** *Fatigue Design*, ESIS 16 (Edited by J. Solin, G. Marquis, A. Siljander, and S. Sipilä) 1993, Mechanical Engineering Publications, London, pp. 171–186.

ABSTRACT Various methods proposed for simplified elastic–plastic analysis of notched components based on linearly elastic solutions are reviewed in the present paper. Elastic and elastic–plastic finite element analyses are carried out for several three-dimensional configurations: (a) a plate of variable thickness with a circular 90 degree through-hole under uniform, uniaxial tension; (b) a thick plate with a circular 90 degree through-hole under bending; and (c) a thick plate with a circular 45 degree through-hole under uniform uniaxial tension. The application of computational results to the prediction of fatigue life of notched components is demonstrated.

Introduction

The need for precise lifetime prediction methods in the design, condition assessment, and failure analysis of mechanical components is widely recognized. The first step in lifetime analysis is to identify the relevant damage and failure modes based on available knowledge of component geometry, loading conditions, and material behaviour.

The most important damage mode under variable mechanical loading is fatigue. As long as load fluctuations are sufficiently small, plastic strains can be neglected and fatigue assessment can be performed on the basis of the linearly elastic stress analysis using 'classical' methods, as shown in engineering handbooks and standards. However, there are many important applications where considerable reversed plastic strains occur. It is then necessary to make a complete elastic–plastic analysis, taking cyclic loading into account. Even with modern computing tools, such an analysis tends to be extremely demanding with respect to computing time and storage required. From this it follows, therefore, that there will always be a need for simplified methods of determining elastic–plastic strain. In particular, this is true for highly localized strain, as it occurs at geometrical irregularities such as holes, grooves, transitions etc., referred to as notches in the following.

Simplified notch analysis

Assuming the material of a notched component to be linearly elastic, the state of stress and strain at the root of the notch is given by σ_{ij}^* and ε_{ij}^*. If the actual

* ABB Power-Generation, Baden, Switzerland.
† The Norwegian Institute of Technology, Trondheim, Norway.

material is elastic–plastic, the real stresses and strains are σ_{ij} and ε_{ij}. One may now ask, whether it is possible to find a relationship between the two sets of stresses and strains.

It may be argued that, as long as gross plasticity does not occur, the strain at the notch root is kinematically controlled by the elastic strain field at some distance from the notch. Thus, Langer (1) assumed the true strain ε_{ij} to be proportional to the elastically calculated, 'fictitious' stress σ_{ij}^*, which implies

$$\varepsilon_{ij} = \varepsilon_{ij}^* \tag{1}$$

This assumption of 'strain invariance' has been shown to provide a useful approximation in cases where strains are mainly thermally induced (2). Based on cyclic strain measurements on notched specimens by Kotani *et al.* (3), it has been suggested by Fuchs and Stephens (4) that equation (1) is also valid under plane strain conditions, whereas for plane stress Neuber's rule predicts plastic strain more accurately. This 'rule' was originally derived by Neuber for a notched prismatic body under shear loading (5), and can be written as

$$\tau \cdot \gamma = \tau^* \cdot \gamma^* \tag{2}$$

In subsequent work, it has been assumed that equation (2) can be extended to a general, multiaxial state of stress and strain through

$$\sigma_e \cdot \varepsilon_e = \sigma_e^* \cdot \varepsilon_e^* \tag{3}$$

where equivalent stress and strain are defined by (2)

$$\sigma_e = (1/\sqrt{2})\{(\sigma_x - \sigma_y)^2 + (\sigma_y - \sigma_z)^2 + (\sigma_z - \sigma_x)^2$$
$$+ 6(\tau_{xy}^2 + \tau_{yz}^2 + \tau_{zx}^2)\}^{1/2} \tag{4}$$

$$\varepsilon_e = (\sqrt{2}/3)\{(\varepsilon_x - \varepsilon_y)^2 + (\varepsilon_y - \varepsilon_z)^2 + (\varepsilon_z - \varepsilon_x)^2$$
$$+ 6(\varepsilon_{xy}^2 + \varepsilon_{yz}^2 + \varepsilon_{zx}^2)\}^{1/2} \tag{5}$$

For a notch in a thin plate, the Neuber rule is often formulated as

$$\sigma_1 \cdot \varepsilon_1 = \sigma_1^* \cdot \varepsilon_1^* \tag{6}$$

where σ_1 and ε_1 denote the maximum principal stress and strain (6). Assuming the state of stress at the notch root to be uniaxial, it can easily be shown that equations (3) and (6) are not equivalent, unless the material is incompressible, i.e. $\nu = \frac{1}{2}$; cf. equation (15) below.

In the following, the validity of Neuber's rule as formulated in equation (3) will be investigated for some simple three-dimensional configurations. Results from finite element analyses are presented graphically as

$$\varepsilon_e/\varepsilon_e^* = f(\sigma_e^*/\sigma_e) \tag{7}$$

Thus, the Neuber rule is satisfied if $f(\sigma_e^*/\sigma_e) = \sigma_e^*/\sigma_e$, whereas strain invariance would imply $f(\sigma_e^*/\sigma_e) = 1$. The corresponding graphs are straight lines in an

$\varepsilon_e/\varepsilon_e^*$ versus σ_e^*/σ_e diagram – the former with a 45 degree slope, the latter running horizontally.

A modification of Neuber's rule, as suggested by Petrequin et al. (7), is given by

$$f(\sigma_e^*/\sigma_e) = (\sigma_e^*/\sigma_e)^m \tag{8}$$

Neuber's rule is regained by letting $m = 1$, whereas $m = 0$ implies strain invariance or strain control, $m = \infty$ stress control.

Finite element analysis of notched components

In this section, results from elastic–plastic finite element analyses of several three-dimensional configurations are presented and discussed. All calculations have been run on an APOLLO 10 000 work station using the ABAQUS finite element code. Isotropic hardening is assumed, and a Ramberg–Osgood type equation is used to describe the uniaxial stress–strain curve, viz.

$$\varepsilon = \sigma/E + \alpha\varepsilon_0^e(\sigma/\sigma_0)^{1/n} \tag{9}$$

Here, ε_0^e denotes the elastic strain at a given, arbitrary stress σ_0. In particular with $\sigma_0 = \sigma_{0.2}$, the 0.2 percent offset yield stress, one obtains for the plastic strain at this stress $\varepsilon_0^p = \alpha\varepsilon_0^e = 0.002$.

Unless otherwise stated, the following materials data have been used in the calculations

E $= 200$ GPa
$\sigma_{0.2} = 600$ MPa
n $= 0.123$

Straight hole in a thick plate under tension

The first case to be analysed is a large plate with a central through-hole in the shape of a circular cylinder with its axis perpendicular to the plane of the plate. The plate is subject to uniform, uniaxial tension, σ_{nom}, on its horizontal boundaries as shown in Fig. 1.

Because of the threefold symmetry of the problem, only one octant of the plate has to be considered. In the plane of the plate, the element division has been based on that of a plane stress model, which yields a stress distribution in close agreement with Kirsch's analytical solution for an infinite elastic plate with a circular hole. In the thickness direction of the plate, the element division has been made fine enough to give a reasonable resolution of stress and strain gradients under all tension levels considered.

The three-dimensional models use 20 node elements, the plane model uses 8 node elements. In both cases reduced integration has been employed. The

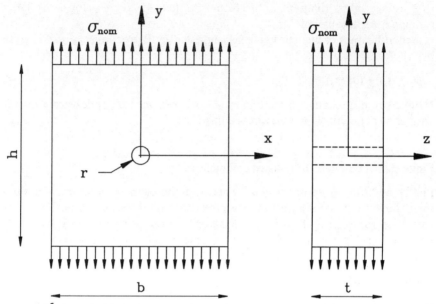

Fig 1   Uniaxially loaded plate with straight circular through-hole ($b = h = 20r$)

mesh used in the analysis of the thickest plate considered ($t = 8r$) is shown in Fig. 2, it consists of 704 elements and has a total of 10 773 degrees of freedom.

A series of analyses has been carried out for plates of increasing thickness t. For $t = 8r$, a state of generalized plane strain is approached in the plane of

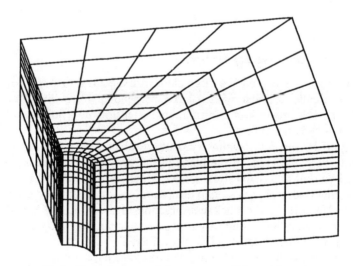

Fig 2   Finite element mesh for one octant of plate with $t = 8r$

symmetry $z = 0$, i.e., a constant strain in the thickness direction

$$\varepsilon_z = \varepsilon_{z\,gps} = -v\sigma_{nom}/E \tag{10}$$

For $t = 0.5r$, a state of plane stress is approached. With $\sigma_z = 0$, one obtains for a linearly elastic plate at $x = r$, $y = 0$, where $\sigma_x = 0$, $\sigma_y = K_{ty}\sigma_{nom}$

$$\varepsilon_z = -v\sigma_y/E = -vK_{ty}\sigma_{nom}/E \tag{11}$$

Normalizing (11) with respect to $|\varepsilon_{z\,gps}|$ according to equation (10) yields

$$\varepsilon_z/|\varepsilon_{z\,gps}| = -K_{ty} \tag{12}$$

These limiting cases are depicted in Fig. 3 together with thickness strain distributions for values of t between $0.5r$ and $8r$.

Under plane stress, plane strain, and generalized plane strain conditions, $K_{ty} = 3.05$. This value is approached at the centre of the plate as $t \ll r$ (plane stress) and $t \gg r$ (generalized plane strain). However, at the plate surface, the stress concentration factor K_{ty} diminishes from the plane stress value 3.05 to approach 2.6 as $t \gg r$. For the plate with $t = 8r$, the longitudinal stress $\sigma_y(z)$ remains almost constant through the thickness to drop sharply just below the plate surface.

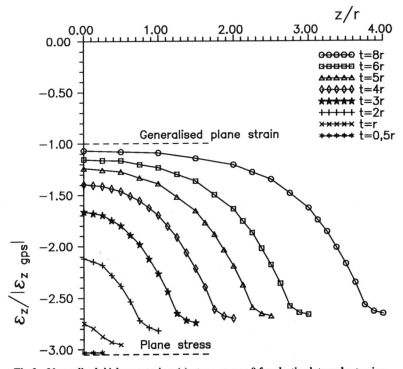

Fig 3   Normalized thickness strain $\varepsilon_z(z)$ at $x = r$, $y = 0$ for elastic plate under tension

A similar through-the-thickness distribution of maximum equivalent stress $\sigma_e(z)$ at the hole radius is observed, as the nominal stress increases sufficiently to yield plastic strains, which are no longer negligible compared to elastic strains. This is shown in Fig. 4 for the thickest plate considered ($t = 8r$). The corresponding distribution of equivalent strain ε_e is shown in Fig. 5, where a sharp strain decrease towards the plate surface can be seen. Figure 5 also illustrates the progressive strain increase with increasing nominal load.

As suggested above the applicability of the Neuber rule to the configuration considered is investigated by plotting $\varepsilon_e/\varepsilon_e^*$ versus σ_e^*/σ_e. With $t = 8r$, the resulting plots for $z = 0$ and $z = t/2$ are given by Fig. 6. Similar results are obtained for $t < 8r$. The curves for $z = 0$ are almost independent of t and agree closely with the nearly coincident curves for plane conditions, i.e., plane stress, plane strain, or generalized plane strain.

Figure 7 demonstrates the pronounced effect of the strain-hardening exponent n on the behaviour of $\varepsilon_e/\varepsilon_e^*$ as a function of σ_e^*/σ_e under generalized plane strain conditions. Since the *ratios* between a pair of strains and stresses are considered, the *levels* of reference strain, $\varepsilon_{0.2} = \sigma_{0.2}/E$, and stress, $\sigma_{0.2}$, become

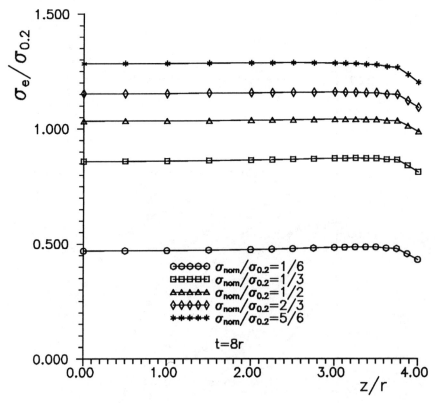

Fig 4   Equivalent stress $\sigma_e(z)$ at $x = r$, $y = 0$ for plate under tension

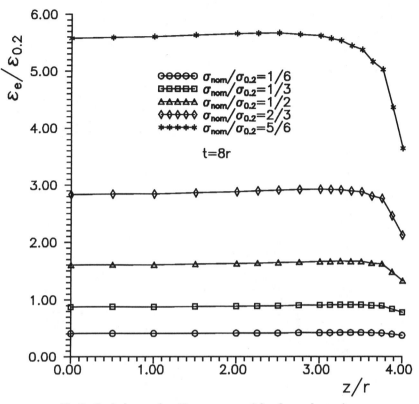

Fig 5   Equivalent strain $\varepsilon_e(z)$ at $x = r$, $y = 0$ for plate under tension

unimportant. Thus the curves depicted in Figs 6, 7, and 13 only depend on the strain hardening exponent n and Poisson's ratio v.

Straight hole in thick plate under bending

Consider the plate with $t = 8r$ subject to out-of-plane bending as shown in Fig. 8.

The same finite element model can be used as in the preceding subsection. The boundary conditions are given by

$$\sigma_y(x, y = h/2, z) = (2z/t)\sigma_{nom}$$

$$u_x(x = 0, y, z) = 0$$

$$u_y(x, y = 0, z) = 0$$

$$u_x(x, y, z = 0) = u_y(x, y, z = 0) = 0$$

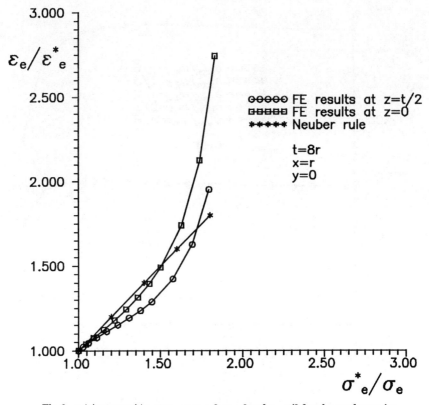

Fig 6 $\varepsilon_e/\varepsilon_e^*$ versus σ_e^*/σ_e at $x = r$, $y = 0$, $z = 0$ and $z = t/2$ for plate under tension

Although nominal bending stress is rising continuously from $z = 0$ to $z = t/2$, the longitudinal stress $\sigma_y(x = r, y = 0, z)$ as well as the equivalent stress $\sigma_e(x = r, y = 0, z)$ have their maximum values just below the plate surface. This phenomenon is closely related to the stress decrease towards the plate surface under uniform tension observed in Fig. 4. As demonstrated in Fig. 9, the maximum equivalent strain also occurs just below the surface of the plate. A graph of $\varepsilon_e/\varepsilon_e^*$ versus σ_e^*/σ_e is shown in Fig. 13.

Oblique hole in a thick plate under tension

The configuration of the straight hole in a thick plate under tension is modified by letting the axis of the circular cylinder be at an angle of 45 degrees with the z axis. In order to simplify the finite element modelling, the plate edges are chosen to be parallel to the cylinder axis as shown in Fig. 10.

Due to symmetry with respect to the plane $y = 0$ and (anti-)symmetry conditions for the plane $z = 0$, it is sufficient to consider a quarter of the plate

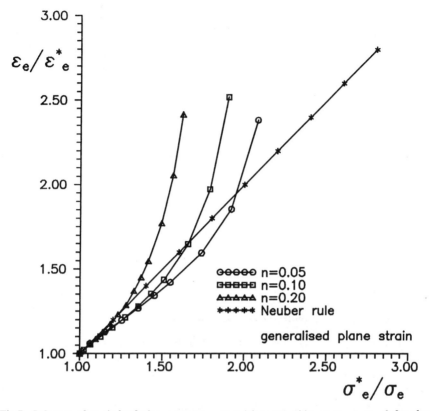

Fig 7 **Influence of strain-hardening exponent n on $\varepsilon_e/\varepsilon_e^*$ versus σ_e^*/σ_e at $x = r$, $y = 0$ for plate under generalized plane strain and uniaxial tension**

defined by $y \geqslant 0$ and $z \geqslant 0$, where the following boundary conditions are imposed

$$u_y(x, y = 0, z) = 0$$

$$u_x(x, y, z = 0) = -u_x(-x, y, z = 0)$$

$$u_y(x, y, z = 0) = u_y(-x, y, z = 0)$$

$$u_z(x, y, z = 0) = -u_z(-x, y, z = 0)$$

The finite element mesh is based on the one used in the preceding sections. However, instead of 20 node volume elements with reduced integration, 15 node triangular prism elements with full integration have been used. The mesh shown in Fig. 11 has 2464 elements and 22 203 degrees of freedom.

Due to the elliptic shape of the intersection between the cylinder and the plate surface, stresses and strains are expected to have their maximum values at the surface. This is confirmed for the equivalent strain by the thickness distributions depicted in Fig. 12.

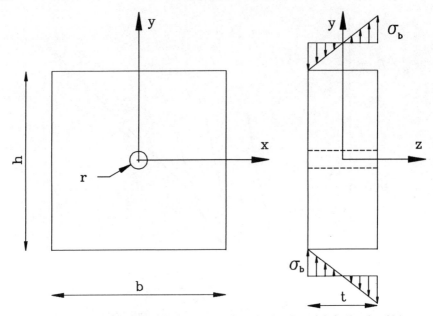

Fig 8 Out-of-plane bending of plate with straight circular through-hole ($b = h = 20r$)

In Fig. 13, $\varepsilon_e/\varepsilon_e^*$ at the surface of the plate has been plotted as a function of σ_e^*/σ_e for the plate with the oblique hole as well as for the plate with a straight hole under bending and tension, respectively. From the appearance of the resulting curves, it is clear that they cannot be described by the power relationship of equation (8); see also Figs 6 and 7.

Fatigue life assessment of notched components

Consider a notched component subject to an alternating stress of constant amplitude. Assuming the Neuber rule to apply, the procedure for predicting the fatigue life of the notched component has been treated by several authors, among others Fuchs and Stephens (4) and Härkegård (8). It will be shown in the following that fatigue life can be determined by adopting the more general relationship (7) between real and fictitious stress and strain.

For equation (7) to apply to cyclic loading conditions, equivalent stress and strain, σ_e and ε_e, must be substituted by equivalent amplitudes, σ_a and ε_a. These are computed from equations (4) and (5) by introducing the *amplitudes* of stress and strain components taking the appropriate signs into account. Furthermore, the monotonic stress–strain curve must be substituted by the cyclic curve. In particular, the Ramberg–Osgood relationship, equation (9), is modified to read

$$\varepsilon_{1a} = \sigma_{1a}/E + 0.002(\sigma_{1a}/\sigma'_{0.2})^{1/n'} \tag{13}$$

For clarity, suffix 1 had been used to emphasize that equation (13) refers to

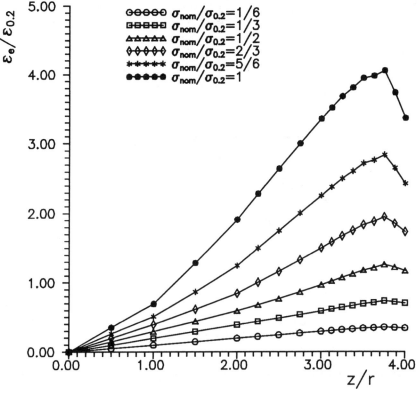

Fig 9 Equivalent strain $\varepsilon_e(z)$ at $x = r$, $y = 0$ for plate under bending

uniaxial (cyclic) loading. $\sigma'_{0.2}$ denotes the 0.2 percent cyclic yield stress, n' the cyclic strain hardening exponent.

Low-cycle-fatigue data, obtained with uniaxially loaded specimens, are often written as **(4)**

$$\varepsilon_{1a} = (\sigma'_f/E)(2N)^b + \varepsilon'_f(2N)^c \tag{14}$$

where the first term of the right hand side equals the elastic strain amplitude ($=\varepsilon^e_{1a}$), and the second term the plastic strain amplitude ($=\varepsilon^p_{1a}$). The equivalent strain amplitude under uniaxial stress becomes

$$\varepsilon_a = \sigma_{1a}/3G + \varepsilon^p_{1a} = (E/3G)\varepsilon^e_{1a} + \varepsilon^p_{1a} \tag{15}$$

Thus, if the equivalent strain amplitude is used to predict fatigue life, the 'elastic' term of equation (14) should be multiplied by the factor $E/3G$ in order to be compatible with equation (15) in the particular case of uniaxial stress. This modification has already been suggested by Manson **(9)** in his classical work on low-cycle fatigue **(9)** and yields the strain–life relation

$$\varepsilon_a = (\sigma'_f/3G)(2N)^b + \varepsilon'_f(2N)^c \tag{16}$$

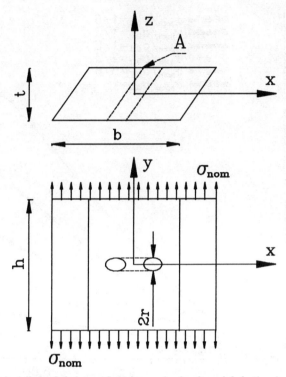

Fig 10   Uniaxially loaded plate with 45 degree circular through-hole ($b = h = 20r$, $t = 8r$)

Combining the total equivalent strain amplitude

$$\varepsilon_a = \sigma_a/3G + \varepsilon_a^p \tag{17}$$

the equivalent stress amplitude

$$\sigma_a = \sigma_f'(2N)^b \tag{18}$$

and equation (16) with the modified Neuber rule, equation (8), yields a relation between fatigue life and the fictitious stress amplitude

$$\sigma_a^* = \sigma_f'(2N)^b\{1 + (3G\varepsilon_f'/\sigma_f')(2N)^{c-b}\}^{1/(1+m)} \tag{19}$$

σ_a^* versus N curves for $m = 0$ (strain control) and $m = 1$ (Neuber rule) are shown in Fig. 14 together with the curve for the plate under bending analysed above.

Conclusions

For a plate under tension or bending with a cylindrical hole of circular section perpendicular to the plane of the plate, the maximum stress and strain are obtained at some distance below the surface. This is in agreement with observations on the location of fatigue cracks in notched components of similar

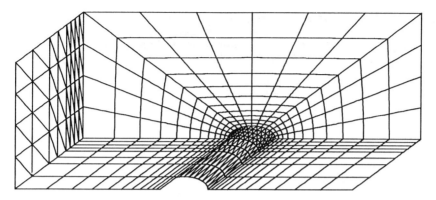

Fig 11   Finite element mesh for one quarter of plate with 45 degree hole

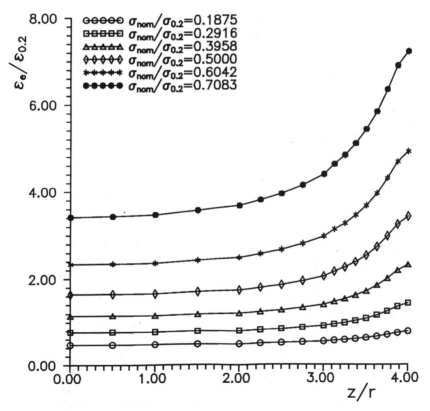

Fig 12   Equivalent strain $\varepsilon_e(z)$ at $x = -\sqrt{2}r + z$, $y = 0$ for plate with 45 degree hole

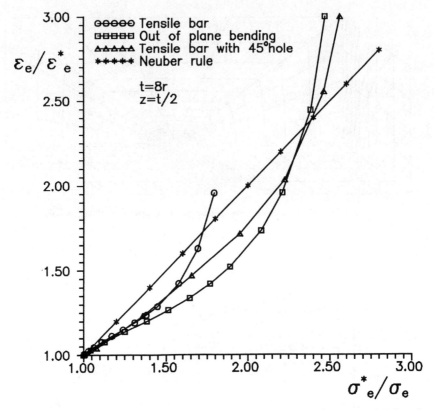

Fig 13 $\varepsilon_e/\varepsilon_e^*$ versus σ_e^*/σ_e at surface of plate with 45 degree hole and plate with through-hole under tension and bending

geometry. Only for an oblique hole, maximum stress and strain occur at the plate surface (acute corner).

For each of these configurations, the validity of the Neuber rule has been investigated by plotting $\varepsilon_e/\varepsilon_e^* = f(\sigma_e^*/\sigma_e)$. It is concluded that the Neuber rule, i.e., $f = \sigma_e^*/\sigma_e$, can be no more than a first approximation of f. In all cases considered in this investigation, the Neuber rule tends to yield conservative estimates of ε_e as long as σ_e^*/σ_e does not exceed a certain value, say 1.5. Based on the present investigation, the contention that notch strains are governed by strain invariance under plane strain conditions must be rejected; in fact, the function f is practically identical for plane stress and (generalized) plane strain.

If more precise strain predictions are needed, it is necessary to carry out a full elastic–plastic analysis. The present work points out some common features of such solutions as well as some peculiarities. Thus, for a given configuration, the function f only depends on the strain-hardening exponent n of a Ramberg–Osgood material.

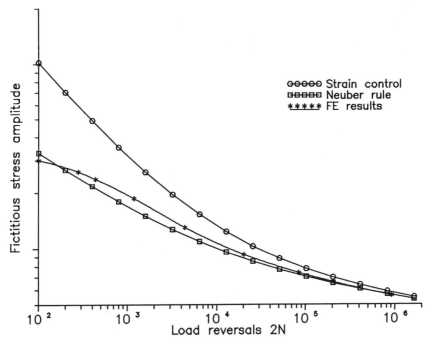

Fig 14 Fictitious stress versus fatigue life for plate with hole under bending. Comparison with idealized assumptions

The function f can be used to determine the relationship between fatigue life and the fictitious, elastically calculated stress amplitude.

Acknowledgements

Our thanks are due to S. Sørbø for substantial support of the finite element analysis, B. Skallerud for valuable discussions on plasticity, and T. Iveland for finalizing the diagrams.

References

(1) LANGER, B. F. (1962) Design of pressure vessels for low-cycle fatigue, *J. basic Engng, Am. Soc. Mech.* **88**, 389–402.
(2) MENDELSON, A. (1968) *Plasticity: theory and application,* McMillan, New York.
(3) KOTANI, S., KOIBUCHI, K., and KASAI, K. (1976) The effect of notches on cyclic stress–strain behaviour and fatigue crack initiation, Proceedings of The Second International Conference on the Mechanical Behaviour of Materials, pp. 606–610.
(4) FUCHS, H. O. and STEPHENS, R. I. (1980) *Metal fatigue in engineering,* John Wiley, New York.
(5) NEUBER, H. (1961) Theory of stress concentration for shear-strained prismatical bodies with arbitrary nonlinear stress strain law, *J. applied Mech.,* **28**, 544–550.
(6) TOPPER, T. H., WETZEL, R. M., and MORROW, JoDEAN (1969) Neuber's rule applied to fatigue of notched specimens, *J. Mater.* **4**, 200–209.

(7) PETREQUIN, P., ROCHE, R., and TORTEL, J. (1983) Life prediction in low cycle fatigue using elastic analysis, *Proceedings of an International Conference on Advances in Life Prediction Methods*, American Society of Mechanical Engineers.

(8) HÄRKEGÅRD, G. (1984) Low-cycle-fatigue analysis of notched components with application to turbine blade fixations, Fatigue 84, Proceedings of The Second International Conference on Fatigue and Fatigue Thresholds, Vol. III., pp. 1659–1668.

(9) MANSON, S. S. (1966) *Thermal stresses and low-cycle fatigue*, McGraw-Hill, New York.

J. R. Sorem* and S. M. Tipton*

The use of Finite Element Codes for Cyclic Stress–Strain Analysis

REFERENCE Sorem, J. R. and Tipton, S. M., **The use of finite element codes for cyclic stress–strain analysis,** *Fatigue Design*, ESIS 16 (Edited by J. Solin, G. Marquis, A. Siljander, and S. Sipilä) 1993, Mechanical Engineering Publications, London, pp. 187–200.

ABSTRACT Commercial finite element codes are increasingly being used to describe elastic–plastic states of stress–strain at structural discontinuities. This paper demonstrates several pitfalls that may be encountered when such codes are utilized for *cyclic* loading situations. A number of simple cyclic loading cases are analysed to demonstrate the conflicting results that can ensue, based on which plasticity option is utilized.

Other factors are addressed that can contribute to analytical uncertainty pertaining to the Bauschinger effect and the use of monotonic versus cyclic constitutive relations.

Introduction

Most mechanical components contain structural discontinuities (notches) that result in localized regions of stress concentration. In order to conduct fatigue analyses of notched components it is often necessary to accurately characterize cyclic states of stress and strain where considerable localized plasticity can occur. The detailed evaluation of stress–strain at a notch is a formidable problem. Since most geometries do not lend themselves to closed form solutions, useful notch strain estimation procedures have been proposed (1)–(3). For a more refined analysis, powerful commercial finite element codes are commonly utilized by engineers to analyze elastic–plastic stress–strain states in complex geometries. However, the application of these codes for cyclic loading is not straightforward.

Often, two or more plasticity relations are embedded in commercial codes (for instance, 'isotropic and kinematic'). These options may give similar or identical results for proportional, monotonic loading but can give drastically different results for a load reversal or for non-proportional loading. Furthermore, the relations are simplified to the point where the material's actual behaviour cannot be accurately portrayed by either option without considerable effort. Examples in the literature of intricate finite element applications to cyclic loading associated with autofrettage are found in references (4)–(6).

In this paper, results from finite element analyses are used to demonstrate problems that can be encountered when modelling elastic–plastic cyclic notch stress–strain states. Fully reversed axial load cycles were applied to finite element models of circumferentially grooved shafts. Three different notch

* Department of Mechanical Engineering, University of Tulsa, Tulsa, Oklahoma, USA.

geometries were investigated, with elastic stress concentration factors of 1.41, 2.54, and 3.49. Material properties were modelled bilinearly with an elastic modulus of 205,000 MPa, a 'yield strength' of 615 MPa, and three different values of tangent modulus, 0, 6000 MPa, and 24 000 MPa. In each case, a maximum applied nominal stress value was estimated using Neuber's rule (1) to produce a notch root strain of 0.01. The maximum load was applied in tension and released, reapplied in compression and released (with a sufficient number of intermediate load steps to achieve convergence within ten iterations), constituting a fully reversed load cycle. The analysis was conducted using the ANSYS® finite element code with the standard kinematic and isotropic hardening options for each geometry and material.

The notch root stress–strain responses predicted by each approach are presented and severe discrepancies noted. Techniques are discussed that enable the use of either hardening rule to model the Bauschinger effect. This discussion is related to the use of monotonic versus cyclic input data.

Hardening rules

The ANSYS® finite element code is capable of using either a kinematic or isotropic hardening rule. Both of these models utilize von Mises yield surfaces. Figure 1 depicts yield surfaces on axes of principal surface stress. A pure kinematic hardening rule assumes that the yield surface maintains its size and shape, translating in stress space in accordance with the stress history. A pure isotropic yield surface maintains its geometric shape in stress space, but changes size to accommodate an expanding stress state.

As shown in Fig. 1, both hardening rules predict identical strain responses from the application of a uniaxial stress, σ_1. While the state of surface stress in an axially loaded notched shaft is biaxial, a uniaxial discussion is presented for sake of illustration. The elastic modulus, E defines the stress–strain response up to a 'yield stress' value of S_y. As the load is increased to σ_1, stress and strain are related by the tangent modulus, E_t.

During unloading, stress and strain are elastically related until the yield surface is again encountered. As shown in Fig. 1, a kinematic hardening rule predicts that reversed yielding will occur once the stress has changed by the amount $2S_y$. However, the isotropic rule predicts that reversed yielding will not occur until a compressive stress equal to $-\sigma_1$ is reached. This fundamental difference between the two approaches is responsible for substantial discrepancies when each are used to model cyclic stress–strain response in an identical finite element geometry.

Finite element geometries

Three different specimen geometries were investigated, as depicted in Fig. 2. Since geometries and loadings were axisymmetric, the models were con-

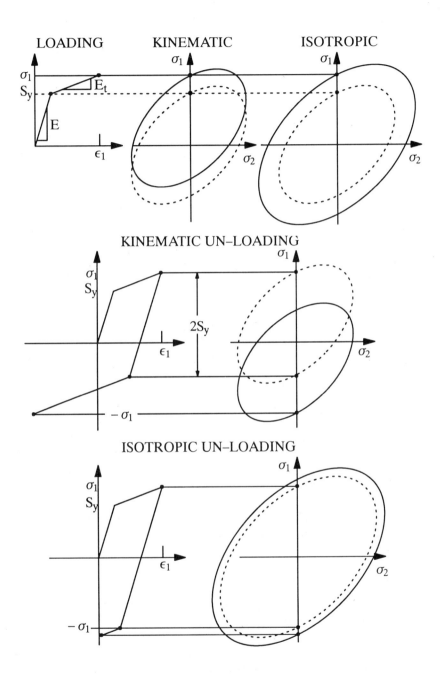

Fig 1 Schematic example of kinematic and isotropic loading and unloading

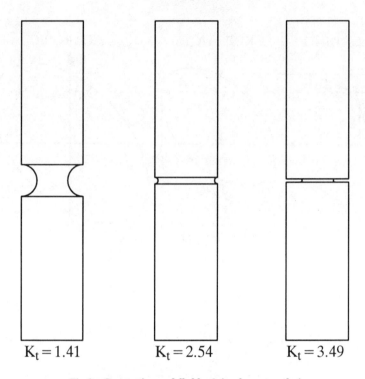

$$K_t = 1.41 \qquad K_t = 2.54 \qquad K_t = 3.49$$

Fig 2   Geometries modelled by finite element analysis

structed using axisymmetric eight-node isoparametric elements. The minimum element dimension (at the critical stress point) ranged from 0.11 percent of the net section diameter for the lowest stress concentration factor model to 0.074 percent of the net section diameter for the two higher stress concentration models. The extreme refinement of the mesh is depicted in Fig. 3. The meshes were refined well beyond the point that would ordinarily be associated with numerical convergence. This was done in order to obtain accurate elastic–plastic strain estimates near the surface for more reliable comparisons of the material models investigated.

Finite element results

The elastic stress concentration factors found for the three specimens are labelled in Fig. 2. For the remainder of the paper, specimens will be referred to as specimen number 1 ($K_t = 1.41$), 2 ($K_t = 2.54$), and 3 ($K_t = 3.49$). The stress concentration factors were 2–5 percent higher than values published by Neuber **(7)** and within 3 percent of those published by Kato **(8)**. The results for specimens number 1 and 3 are approximately 14 percent and 31 percent higher, respectively, than the values suggested by Peterson **(9)**.

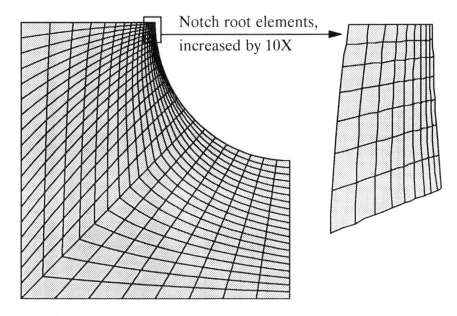

Fig 3   Fine grid used to model notch roots

A total of 18 finite element runs are considered in this paper. Tables 1–3 provide an overview of results for each of the three geometries. Listed in the tables are the applied nominal stresses and the multiaxial notch stress–strain response predicted at the reversal points.

Discussion

From the results in Tables 1–3, it is apparent that considerable differences resulted from the use of isotropic versus kinematic hardening. The kinematic results are presented graphically in Figs 4–6 along with predictions made by a simple application of Neuber's Rule (10)–(12) that approximates the state of stress as purely uniaxial and a Massing (13) material (as predicted by kinematic hardening). The nominal load levels for the finite element models were selected to result in identical Neuber notch stress–strain estimates for all three geometries. Neuber estimates are shown for the first half cycle (load–unload) and the complete cycle (load–compression–unload).

Notice in Figs 4–6 that the load–unload portion of the cycle results in a residual compressive stress at the notch root. The simple Neuber's rule estimated compressive stresses of -508, -504, and -455 MPa for the specimens 1–3, respectively, for all three tangent moduli. The kinematic finite element results (in Tables 1–3) agree with the Neuber estimates in that the three unloading residual compressive stress estimates are within 5 percent for a

Table 1 Loading and notch stress–strain response for $K_t = 1.41$

Tangent modulus	$E_t = 0$		6000 MPa		24 000 MPa	
Hardening model	KIN	ISO	KIN	ISO	KIN	ISO
Loading						
S_{nom} (max) (MPa)	796	796	823	823	898	898
S_{nom} (min) (MPa)	−796	−796	−823	−823	898	−898
Notch multiaxial stress–strain response						
σ_z (max) (MPa)	677	677	705	705	835	834
σ_θ (max) (MPa)	143	142	137	136	152	150
ε_z (max) (x 10^{-6})	7837	7834	8357	8349	9417	9414
ε_θ (max) (x 10^{-6})	−1665	−1666	−1867	−1865	−2186	−2185
ε_r (max) (x 10^{-6})	−1284	−1282	−1323	−1322	−1543	−1539
σ_z (zero) (MPa)	−442	−442	−449	−450	−427	−427
σ_θ (zero) (MPa)	−63	−64	−75	−76	−80	−82
ε_z (zero) (x 10^{-6})	2683	2681	3034	3027	3603	3600
ε_θ (zero) (x 10^{-6})	−1034	−1035	−1215	−1213	−1474	−1473
ε_r (zero) (x 10^{-6})	650	652	673	.675	638	643
σ_z (min) (MPa)	−670	−675	−697	−756	−827	−1010
σ_θ (min) (MPa)	−123	−128	−112	−170	−129	−279
ε_z (min) (x 10^{-6})	−7864	−7789	−8373	−7286	−9447	−5338
ε_θ (min) (x 10^{-6})	1628	1579	1399	1187	687	180
ε_r (min) (x 10^{-6})	1246	1259	1276	1424	2978	1898
σ_z (zero) (MPa)	448	443	458	399	434	252
σ_θ (zero) (MPa)	83	78	101	43	104	−46
ε_z (zero) (x 10^{-6})	−2709	−2635	−3047	−1963	−3632	476
ε_θ (zero) (x 10^{-6})	997	948	1173	535	1454	−531
ε_r (zero) (x 10^{-6})	−688	−674	−721	−573	−683	−283

given specimen. However, the use of Neuber's rule overpredicted the residual compressive stresses from 12 to 22 percent.

The multiaxial stress state at the notch root is apparent in Tables 1–3. This has the effect of causing the curves in Figs 4–6 from the Neuber and finite element analyses not to be coincident. In these figures, the axial direction stress is being plotted from the finite element results while the simplified application of Neuber's rule assumes a uniaxial stress state. The notch constraint effect increases with the severity of the notch since the Neuber's over-approximation of notch strain increases with the stress concentration factor. It is possible to refine the Neuber analysis to account for notch root multi-axiality (3), but this was beyond the scope of the current investigation.

Tables 1 and 2 indicate that the isotropic hardening model is entirely inappropriate for cyclic loading without substantial modification. Experience has shown that fully-reversed loading of notched components usually results in a fully-reversed stress–strain response. The kinematic finite element results agree with this observation, but the isotropic results are extremely asymmetric, as expected. In fact, the ANSYS® user's manual recommends against the use of isotropic hardening for cyclic loading, but such warnings are not generally obvious.

Table 2 Loading and notch stress–strain response for $K_t = 2.54$

Tangent modulus	$E_t = 0$		6000 MPa		24 000 MPa	
Hardening model	KIN	ISO	KIN	ISO	KIN	ISO
Loading						
S_{nom} (max) (MPa)	442	442	457	457	499	499
S_{nom} (min) (MPa)	−442	−442	−457	−457	−499	−499
Notch multiaxial stress–strain response						
σ_z (max) (MPa)	710	734	825	710	733	826
σ_θ (max) (MPa)	276	285	311	276	286	314
ε_z (max) (x 10^{-6})	6405	6640	7178	6402	6638	7174
ε_θ (max) (x 10^{-6})	−430	−446	−491	−430	−446	−491
ε_r (max) (x 10^{-6})	−1581	−1636	−1810	−1581	−1634	−1813
σ_z (zero) (MPa)	−413	−427	−443	−413	−427	−443
σ_θ (zero) (MPa)	19	19	21	20	20	24
ε_z (zero) (x 10^{-6})	1311	1372	1424	1309	1371	1421
ε_θ (zero) (x 10^{-6})	−25	−28	−34	−25	−28	−34
ε_r (zero) (x 10^{-6})	398	412	426	397	411	423
σ_z (min) (MPa)	−712	−736	−827	−717	−785	−1002
σ_θ (min) (MPa)	−305	−312	−335	−306	−324	−365
ε_z (min) (x 10^{-6})	−6350	−6593	−7163	−6333	−6325	−6265
ε_θ (min) (x 10^{-6})	427	444	490	427	437	465
ε_r (min) (x 10^{-6})	1618	1674	1845	1625	1734	2050
σ_z (zero) (MPa)	410	425	444	406	376	266
σ_θ (zero) (MPa)	−48	−47	−44	−49	−58	−75
ε_z (zero) (x 10^{-6})	−1260	−1324	−1398	−1241	−1057	−511
ε_θ (zero) (x 10^{-6})	23	26	33	23	19	8
ε_r (zero) (x 10^{-6})	−357	−373	−396	−353	−314	−187

The bilinear approximation of material stress–strain behaviour could be considered overly restrictive when modelling materials with higher strain hardening exponents. The use of a bilinear material characterization was made in this investigation for the sake of simplifying the overall example. It is possible in most finite element codes to utilize a multi-linear material characterization when additional refinement is necessary.

Notice in Figs 4–6 that the Massing (13) behaviour predicted by kinematic hardening causes the magnitude of the compressive reversed yield stress to be smaller than the tensile yield stress for the two strain hardening materials investigated ($E_t = 6000$ and 24 000 MPa). This observation is consistent with the Bauschinger effect (14). However, cyclic material data often reveal a substantially different unloading response than that predicted by a kinematic rule (4). Figure 7 shows a stress–strain loading and reverse loading curve (first three-quarter cycle) from a low cycle fatigue test of a specimen removed from a 4340 forging. The loading portion of the curve is the monotonic stress–strain curve. The cyclic stress–strain curve is shown on the same figure. For kinematic unloading, the stress–strain response is predicted by geometrically doubling the monotonic or cyclic curves. These estimated unloading curves are

Table 3 Loading and notch stress–strain response for $K_t = 3.49$

Tangent modulus	$E_t = 0$		6000 MPa		24 000 MPa	
Hardening model	KIN	ISO	KIN	ISO	KIN	ISO
Loading						
S_{nom} (max) (MPa)	321	321	333	333	363	363
S_{nom} (min) (MPa)	−321	−321	−333	−333	−363	−363
Notch multiaxial stress–strain response						
σ_z (max) (MPa)	714	735	818	714	735	818
σ_θ (max) (MPa)	311	321	351	311	321	353
ε_z (max) (x 10^{-6})	5817	6057	6580	5812	6058	6578
ε_θ (max) (x 10^{-6})	49	48	48	49	48	48
ε_r (max) (x 10^{-6})	−1615	−1668	−1838	−1616	−1669	−1842
σ_z (zero) (MPa)	−388	−405	−425	−389	−405	−425
σ_θ (zero) (MPa)	−36	−38	−40	−36	−38	−38
ε_z (zero) (x 10^{-6})	957	1028	1099	951	1030	1097
ε_θ (zero) (x 10^{-6})	−21	−24	−30	−21	−24	−30
ε_r (zero) (x 10^{-6})	464	485	508	466	484	505
σ_z (min) (MPa)	−713	−736	−819	−718	−779	−979
σ_θ (min) (MPa)	−333	−342	−370	−334	−355	−413
ε_z (min) (x 10^{-6})	−5779	−6005	−6594	−5770	−5821	−5899
ε_θ (min) (x 10^{-6})	−50	−49	−49	−50	−54	−67
ε_r (min) (x 10^{-6})	1634	1693	1868	1643	1754	2082
σ_z (zero) (MPa)	388	401	428	384	358	264
σ_θ (zero) (MPa)	15	17	22	13	3	−21
ε_z (zero) (x 10^{-6})	−928	−993	−1098	−918	−808	−419
ε_θ (zero) (x 10^{-6})	20	22	30	20	18	11
ε_r (zero) (x 10^{-6})	−435	−451	−485	−429	−392	−265

superimposed with the actual unloading curve in Fig. 7. This emphasizes the fact that the unloading curve can be substantially different from the kinematic representation of the loading curve. It has been suggested (15) that a kinematic (Massing) implementation of the cyclic stress–strain curve provides a better representation of a material's path-dependent incremental constitutive behaviour. Therefore, for constant amplitude cyclic loading, the use of the cyclic stress–strain curve would be recommended. However, for situations where the first load application is intentionally higher than subsequent service loading, it may be necessary to utilize the monotonic curve for the initial load application and then to change the material model for the unloading and subsequent service loading (4)–(6). The first step overload often occurs when autofrettage or proof loading is used to intentionally impose residual compressive stress in components subjected to $R = 0$ (zero-to-maximum) loading, such as pressure vessels, lifting components, and tackle.

In view of the above discussion, it is possible to make adjustments to either kinematic, isotropic, or a combination of the two hardening rules to account for the Bauschinger effect during the first load reversal, before changing to the material's cyclic stress–strain relation. Two examples are depicted in Fig. 8

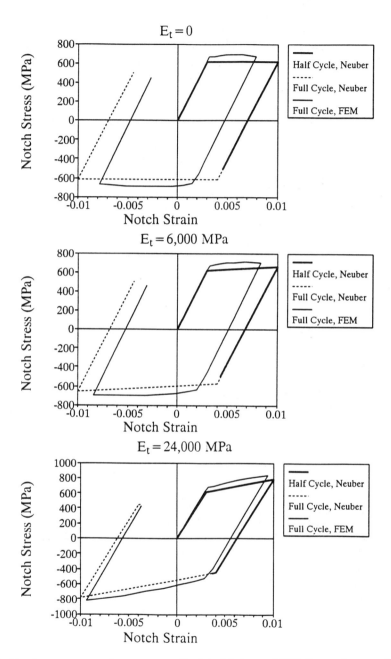

Fig 4   Notch root stress–strain response predicted by Neuber and FEA for specimen number 1
($K_t = 1.41$)

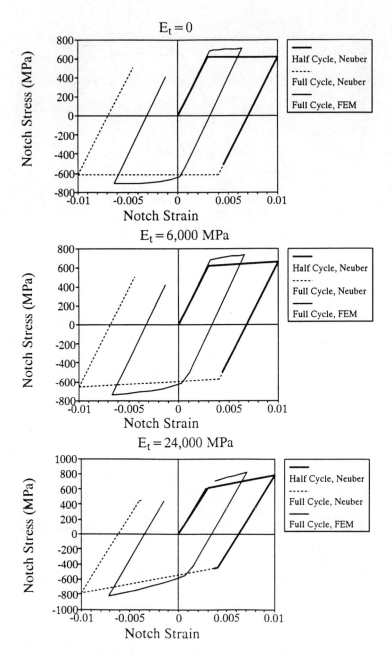

Fig 5 **Notch root stress–strain response predicted by Neuber and FEA for specimen number 2**
($K_t = 2.54$)

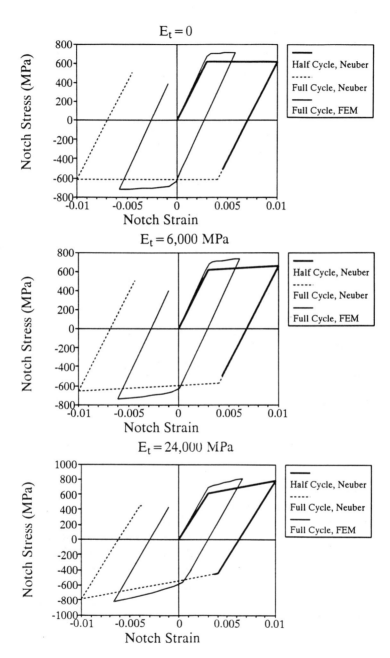

Fig 6 Notch root stress–strain response predicted by Neuber and FEA for specimen number 3 ($K_t = 3.49$)

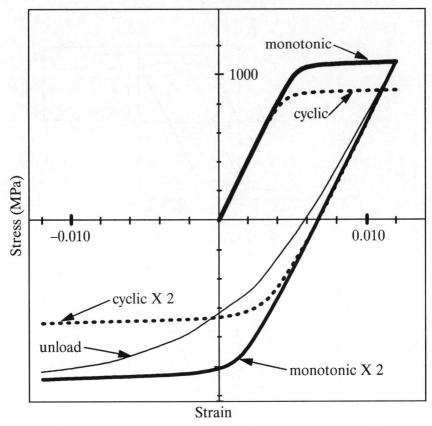

Fig 7 **First three-quarter cycle for 4340 forged steel specimen, superimposed with kinematic monotonic and cyclic unloading curve. (Kinematic unloading curves are equivalent to the loading curves, magnified by a factor of two)**

using isotropic hardening and a combination of kinematic and isotropic hardening. It should be kept in mind that these adjustments would be necessary for each individual element in a general finite element analysis. To expedite this procedure, it is possible to categorize elements into 'strain groups' and adjust the properties on a group basis or assign material properties based on temperature. This can be done by assigning material properties based on temperature and adjusting the nodal temperature to modify the element properties. In any event, the procedure is very cumbersome, analytically, and currently must be performed external to the finite element analysis.

Conclusions

There is an increasing emphasis on the use of finite element analysis to characterize states of elastoplastic stress and strain in engineering design applica-

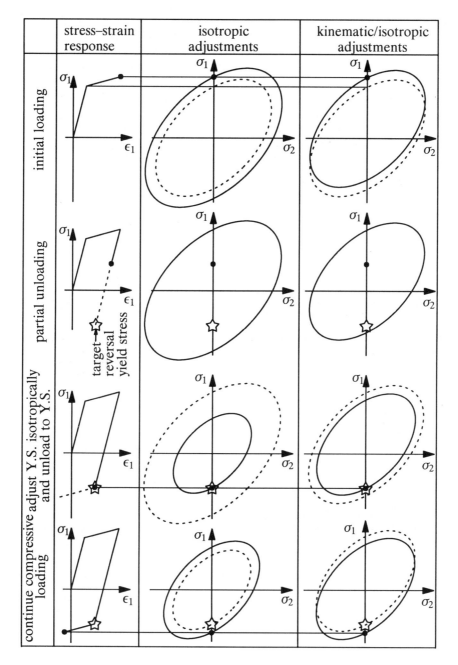

Fig 8 The use of isotropic or modified kinematic/isotropic to model the first load reversal during an overload operation

tions. Thus it is important to realize that cyclic loading cannot be handled straightforwardly with existing commercial finite element codes. The results of this investigation graphically depict why finite element modelling of cyclic loading should be conducted with a kinematic rather than isotropic hardening rule when only these options are available. Kinematic hardening does exhibt a Bauschinger effect, but the accuracy of the unloading and subsequent cyclic material response by a purely kinematic rule will not generally be correct. If more detailed data are available from cyclic material testing, modifications to both kinematic and isotropic hardening models can be made to increase correlation with those data. However, the process is user-intensive and should be incorporated as an option to commercial finite element packages.

References

(1) NEUBER, H. (1961) Theory of stress concentration for shear strained prismatical bodies with arbitrary nonlinear stress–strain law, *J. appl. Mech.*, **12**, 544–550.

(2) MOLSKI, K. and GLINKA, G. (1981) A method of elastic–plastic stress and strain calculation at a notch root, *Mater. Sci. Engng*, **50**, 93–100.

(3) HOFFMANN, M. and SEEGER, T. (1985) A generalized method for estimating multiaxial elastic–plastic notch stresses and strains, Parts 1 and 2, *J. Engng Mater. Technol.*, 250–260.

(4) MILLIGAN, R. V., KOO, W. H., and DAVIDSON, T. E. (1965) (1966) The Bauschinger effect in a high strength steel, Technical Report, Benet Laboratory, US Army, WVT 6508, also *J. bas. Engng*, 480–488.

(5) CHAABAN, A. (1985) *Static and fatigue design of high pressure vessels with blind-ends and cross-bores*, PhD Thesis, University of Waterloo, Ontario, Canada.

(6) KOTTGEN, V. B., SCHON, M., and SEEGER, T. (1991) Application of a multiaxial load-notch strain approximation procedure to autofrettage of pressurized components, ASTM Symposium on Multiaxial Fatigue, to be published.

(7) NEUBER, H. (1987) *Kerbspannunglehre*, Springer, Berlin.

(8) KATO, A. (1991) Design equation for stress concentration factors of notched strips and grooved shafts, *J. Strain Analysis*, **26**, 21–28.

(9) PETERSON (1974) *Stress concentration factors*, Wiley and Sons, New York.

(10) FUCHS, H. O. and STEPHENS, R. I. (1980) Metal fatigue in engineering, Wiley and Sons, New York.

(11) *Fatigue design handbook* (1990) Society of Automotive Engineers, PA.

(12) BANNANTINE, J. A., COMER, J. J., and HANDROCK, J. H. (1990) *Fundamentals of metal fatigue analysis*, Prentice Hall, New Jersey.

(13) MASSING, G. (1926) Eigenspannungen und Versfestigung beim Messing. Proceedings, Second International Congress for Applied Mechanics.

(14) BAUSCHINGER, J. (1881) Über die Veränderung der Elasticitätsgrenze und des Elasticitäts-moduls verschiedener Metalle, *Zivilingenieur*, **27**, 289–348.

(15) DOWLING, N. E., BROSE, W. R., and WILSON, W. K. (1977) Notched member fatigue life predictions by the local strain approach, *Fatigue under complex loading: analyses and experiments*, Society of Automotive Engineering, PA, pp. 55–84.

D. Socie*

Discriminating Experiments for Multiaxial Fatigue Damage Models

REFERENCE Socie, D., **Discriminating experiments for multiaxial fatigue damage models,** *Fatigue Design*, ESIS 16 (Edited by J. Solin, G. Marquis, A. Siljander, and S. Sipilä) 1993, Mechanical Engineering Publications, London, pp. 201–211.

ABSTRACT Multiaxial fatigue damage models and the experiments that are needed to discriminate between the models and determine the material constants are reviewed. Only those models capable of assessing fatigue damage for variable amplitude non-proportional loading histories commonly found in structures are included.

Introduction

Many multiaxial fatigue damage parameters have been proposed, each with a different set of experiments. This paper examines the models and experiments that could be used to confirm or refute them. Fatigue damage is influenced by many factors too numerous to list here. For purposes of estimating the fatigue life of structures and components, models for multiaxial loading should be able to account for at least the following:

material;
state of stress/strain;
mean stress/strain;
nonproportional loading; and
variable amplitude.

This paper reviews some of the models and their ability to meet the requirements listed above. Many of the models were proposed as a result of observations of crack nucleation and growth and it is appropriate to briefly review them.

Fatigue damage is best described as the nucleation and growth of cracks to final failure. In 1903 Ewing and Humfrey **(1)**, motivated by the work of Wohler and Bauschinger, published their classic paper, 'The Fracture of Metals under Repeated Alternations of Stress'. Their description of the fatigue process follows: 'The cracks occurred along broadened slip-bands: in some instances they were first seen on a single crystal, but soon they joined up from crystal to crystal, until finally a long continuous crack was developed across the surface of the specimen. When this happened a few more reversals brought about fracture.' These authors noted that: 'Once an incipient crack begins to form across a certain set of crystals, the effect of further reversals is mainly confined to the neighborhood of the crack.' These slip-lines, more commonly

* Department of Mechanical Engineering, University of Illinois at Urbana-Champaign, Urbana, Illinois, USA.

called persistent slip bands, are caused by the movement of dislocations. The crystals are individual grains in the material. Dislocations move only on their crystallographic slip planes under an applied shear stress. When the critical resolved shear stress in a grain is exceeded, the dislocations move and this results in plastic shear strains.

During tensile loading shear stresses are produced on planes that are oriented at 45 degrees to the tensile axis. Grains whose crystallographic slip planes and directions are also oriented at 45 degrees to the tensile axis will have the highest critical resolved shear stress and plastic strains and will be the first to form slip bands and cracks. Given this description of the process, it is clear that the macroscopic cyclic shear stress and the resulting plastic shear strain are the driving forces for crack nucleation and should be the appropriate parameters for correlating test data for various states of stress such as tension/compression and torsion. Equal cyclic shear stress or strains should result in equivalent fatigue damage. Unfortunately this is not always the case. A more complete understanding requires consideration of how small cracks grow from the slip band that forms in a single grain. In some materials and loading conditions the majority of the fatigue life is consumed in growing small cracks from the order of the grain size to a length of a few millimeters. Hence, their growth is more important than their nucleation. Macroscopic crack growth can occur on a plane perpendicular to the maximum principal stress even though the local growth at the crack tip is a shear strain controlled process. Viewed on a macroscale, that is on a scale larger than the grain size, tensile stresses are also responsible for the growth of fatigue damage and should be an appropriate damage parameter in some instances.

Brown and Miller (2) provide a comprehensive review of the literature in terms of strain. They considered the nucleation and growth of fatigue cracks

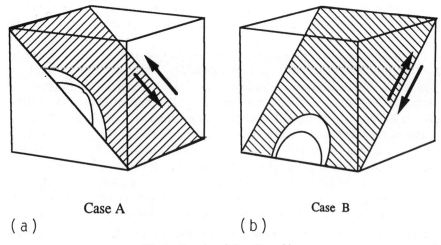

Case A   Case B
(a)   (b)

Fig 1   Case A and Case B cracking

and suggested the terms 'case A' and 'case B' cracks. Case A cracks are illustrated in Fig. 1(a) for torsion loading. In torsion loading, the shear stress is in-plane and acts on the free surface in a direction parallel to the length of the crack. This results in a mode II shear crack on the surface that changes to mode III in the interior. There is no shear stress acting perpendicular to the free surface along the crack depth. Cracks tend to be shallow and have a large aspect ratio. In biaxial tension (case B), the shear stress causes the cracks to grow into the depth as shown in Fig. 1(b). This shear stress is out-of-plane, resulting in a mode II crack in the interior. These cracks have a mode I component and frequently turn and grow as mode I tensile cracks after the crack advances beyond a few grains. Case B cracks are the type described by the intrusion–extrusion model. Tension loading has the same shear stress for both case A and case B and can display either mode of cracking. Combined tension/ torsion loading always has case A cracks.

Damage models

Fatigue damage models can be based on stress, strain, or combinations of stress and strain. The following section briefly reviews representative damage models that have the potential to meet the requirements listed in the introduction. A number of other models exist but it is difficult to see how they can be used in their present formulations for variable amplitude non-proportional loading; they are not included here.

Stress based

Based on physical observations of the orientation of initial fatigue cracks in steel and aluminium, Findley (3) discussed the influence of normal stress acting on the maximum shear stress plane. A critical plane model was introduced (4)

$$\tau_a + k\sigma_{n,\,max} = f(N_f) \tag{1}$$

For a constant fatigue life, the allowable alternating shear stress, τ_a, decreases with an increase in the maximum normal stress, $\sigma_{n,\,max}$, on the plane of the critical alternating shear stress. Here, the maximum normal stress was formulated as the sum of the normal stress resulting from the amplitude and mean stress. An empirical constant k is used to fit the experimental data. The right hand side is a function of the fatigue life, N_f.

McDiarmid (5) conducted an extensive survey of literature on multiaxial fatigue in the high-cycle regime in 1972. He showed that the ellipse quadrant proposed by Gough (6) can be divided into components of maximum shear stress amplitude and the normal stress acting on the plane of maximum shear stress amplitude similar to Findley's model. McDiarmid argued that his proposed model is based on physical observations on the effect of normal stress on the maximum shear stress orientation; whereas the Gough ellipse quadrant is empirical. The most recent formulation of his work (7) can be expressed in the finite life region as

case A $\quad \tau_a(1 + \sigma_{n, max}/2\sigma_u) = f(N_{f, A})$

case B $\quad \tau_a(1 + \sigma_{n, max}/2\sigma_u) = g(N_{f, B})$ $\qquad\qquad$ (2)

This failure criterion is also based on the shear stress amplitude and the maximum normal stress on the plane of maximum shear stress amplitude. The tensile strength is denoted σ_u. The model considers two types of shear cracking modes. Case A cracks growing along the surface have a stress life curve denoted by $f(N_{f, A})$. Case B cracks growing into the surface have a separate set of material constants that govern the fatigue life $g(N_{f, B})$. Case A cracks are found in torsion loading and case B cracks occur under biaxial tension loading so that the proposed model accomodates differences in the fatigue limit in tension and torsion through differences in the functions f and g.

Strain based

Brown and Miller (2) proposed that cracks nucleate on planes of maximum shear strain and suggested that the critical parameters for the nucleation and early growth of fatigue cracks were the maximum shear strain amplitude, γ_a, and the strain amplitude normal to the maximum shear strain plane, ε_n. They were the first to propose separate relationships for case A and case B cracks. A more convenient form of the theory for case A is given by Kandil *et al.* (8)

case A $\quad \gamma_a + S\varepsilon_n = f(N_{f, A})$

case B $\quad \gamma_a = g(N_{f, B})$ $\qquad\qquad$ (3)

Here an empirical constant S, is used to fit the test data for a particular material. In many cases the constant S may be set equal to unity. Socie *et al.* (9) added a mean stress term to this model for case A cracking. Mean stress normal to the maximum shear strain amplitude plane, σ_{no} was normalized by the elastic modulus, E, to preserve the dimensionless feature of strain with the result

$$\gamma_a + \varepsilon_n + \sigma_{no/E} = f(N_f) \qquad\qquad (4)$$

They did not specify case A or case B cracking but conducted all of the experiments in combined tension and torsion where the cracking mode is case A. Later Fatemi and Socie (10) proposed a formulation of the parameter which was more suitable for variable amplitude non-proportional loading and which included the maximum stress normal to the plane of maximum shear strain amplitude

$$\gamma_a(1 + \sigma_{n, max}/\sigma_y) = f(N_f) \qquad\qquad (5)$$

The normal stress was normalized with the yield strength, σ_y. This model has been successfully used for both case A and case B cracks by Morrow (11) and no distinction was made between case A and case B formulations as in the other models.

Thus far only shear models have been included. It is appropriate to consider a tensile strain model for those materials that exhibit mode I cracking as the primary failure mode. One such model that has gained widespread use in uniaxial loading is the Smith *et al.* (12) parameter composed of the normal strain amplitude, ε_a, and maximum stress. For multiaxial loading the maximum stress, σ_{max}, is interpreted as the maximum stress normal to the maximum normal strain amplitude plane (13) resulting in a model in this form

$$\sigma_{max} \varepsilon_a = f(N_f) \tag{6}$$

Given the number of models it is appropriate to ask which experiments are needed to discriminate between the models and to determine the material constants for fatigue life calculations? How will the damage models be used for variable amplitude non-proportional loading?

Discussion

All of the constants for these models could be obtained from uniaxial tension tests with and without mean stresses. Unfortunately the tension test alone is not a discriminating experiment because the failure mode can take several forms. Depending on the material and stress or strain amplitude, the failure mechanism could be mode I tensile cracks, case A shear cracks, or case B shear cracks. Tension tests are appropriate only when the failure mode is known. Torsion tests where case A cracks form should be used to develop the material constants for most of the models. Even in torsion the failure mode must be known since low ductility materials are likely to form mode I tensile cracks. Tension–tension testing is required to determine a model's application for case B shear cracking. This type of testing is difficult experimentally and little test data exists.

Some care must be used in determining the constants for the damage models. All of the damage models listed above were formulated on the basis of simple proportional loading tests. Interpretation of the damage parameter for complex loadings poses some problems. For simple short loading histories such as in and out-of-phase sine waveforms, the maximum value of the strain amplitude and the damage parameter are always on the same plane. For complex loading, the maximum shear strain amplitude and maximum value of the damage parameter may be on different planes. Should damage be calculated on the planes that experience the maximum strain amplitudes or should damage be computed on the basis of the maximum value of the damage parameter? A simple example shows the need to compute damage based on the maximum value of the damage parameter. Consider a few large torsion cycles applied to a tubular specimen followed by a large number of tension cycles with a smaller strain range. The plane experiencing the largest range of shear strain would be at 0 degrees and 90 degrees from the torsion cycles. Failure, however, would be expected to occur on the 45 degree shear planes from all of the tension cycles.

Consider the case of simple torsion loading used to establish the material constants for the shear damage models. Using model 5 as an example, the critical plane is predicted to be 10 degrees rather than 0 degrees and 90 degrees as expected. Mohr's circle for this loading shows that the shear strain on the 10 degree plane is 6 percent less than the maximum. The normal stress on the 10 degree plane is increased from 0 to 35 percent of the maximum stress. As a result the damage parameter reaches a maximum on the 10 degree plane. Material constants must be defined in the same manner as the intended use of the damage model to avoid predicting unrealistic damage. Material constants must be defined in terms of the damage parameter and not in terms of the maximum stress or strain amplitude to be applicable to variable amplitude nonproportional loading histories.

For convenience, fatigue models will be referred to by equation number in this discussion. With this interpretation of the damage parameters, models 1, 3A, and 4 have the potential to predict unrealistic damage for some loading histories. Shear damage would be computed on a plane with no shear stress or strain amplitude but with a normal strain amplitude or static mean stress. As a result, these formulations are not appropriate for variable amplitude nonproportional loading.

A simple set of experiments could be used to distinguish between models based on tensile stress and shear stress. Consider a tubular specimen that can be loaded in tension, torsion, and internal pressure. The baseline test will be torsion. Materials that exhibit extensive shear cracking (case A) are expected to have a distribution of damage on both 0 and 90 degrees planes where the 0 degree plane represents a direction parallel to the tube axis. Adding static tension in the hoop direction would introduce a tensile stress on only one of the two shear planes and would reduce the fatigue life from the case of torsion alone (14). Static hoop stress will not influence the stress on the 90 degree shear plane. The static hoop stress will result in increased stresses on both 45 degree tension planes. These increased stresses and their directions are shown by the arrows in Fig. 2(a). The addition of static hoop stress will be detrimental for both materials that fail in shear and for materials that fail in tension. Now consider the case of static axial compression with cyclic torsion. The 90 degree shear plane will have a compressive stress, but this beneficial compressive stress will not increase the fatigue life of a material that forms case A cracks because the material will fail on the 0 degree shear plane indicated in Fig. 2(b). Both 45 degree tensile planes will see the beneficial effect of the compressive stress. This discriminating test suggests that the compressive stress will not have an influence on the fatigue life for materials that fail in shear and have a large influence for materials that fail in tension. In the next test case, internal pressure and axial compression are used to obtain static hoop tension stress and static compression axial stress which is superimposed on cyclic torsion. Mean stress on the 0 degree shear plane is compressive and mean stress on 90 degree plane is tensile as illustrated in Fig. 2(c). The

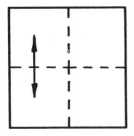 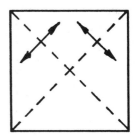

Shear Damage   Tensile Damage

Fig 2(a)   Torsion with static hoop tension

expected fatigue life will be reduced for case A cracking and failure expected
on the 90 degree plane. The two stresses will combine and cancel each other
on the 45 degree tension plane. In this critical test, the shear damage material
is expected to show a significant decrease in fatigue life and the tensile material
should be unaffected.

In the preceding example the torsion stress or strain amplitudes were kept
constant, only the mean stress was varied. All of the damage models listed will
correctly predict the decrease in fatigue life for the first mean stress loading
case with static tension. Model 6 predicts that a static axial compression stress
will increase the fatigue life. The other models predict that the fatigue life will
be the same as torsion alone. In the axial compression with hoop tension test,
model 6 predicts that the life will be the same as torsion alone. The other
models predict this test will have the same decrease in fatigue life as the hoop
tension test. These tests will provide a set of data to confirm or refute the

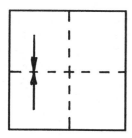 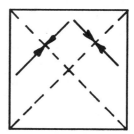

Shear Damage   Tensile Damage

Fig 2(b)   Torsion with static axial compression

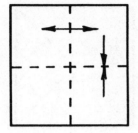   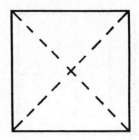

Shear Damage   Tensile Damage

Fig 2(c)   Torsion with static hoop tension and axial compression

suitability of the Smith–Watson–Topper damage parameter; they will not
separate the different shear damage models listed above. Shear models have
been proposed that include a hydrostatic stress term. Given the preceding discussion, these models seem inappropriate in that they will predict that a compressive mean stress on one shear plane will cancel the effects of a tensile mean
stress on another shear plane.

A set of experiments that would clearly discriminate between case A and
case B cracking in tension is difficult. One possibility would be to conduct a
test on a tube that is cyclically loaded in tension with a superimposed static
hoop stress. For case A cracking, the hoop stress will result in a resolved
tensile mean stress that is normal to the crack surface and will be expected to
reduce the fatigue life. The tensile mean stress will be parallel to the crack
surface for case B cracks and not be expected to influence the fatigue life.
Unfortunately, these tests will not differentiate between case B shear cracking
and mode I tensile cracks.

Non-proportional loading involves situations in which the principal stresses
and strain rotate. In the plastic strain region, this rotation can result additional cyclic hardening that is not observed in uniaxial tests **(13)**. Steels which
exhibit cyclic softening under uniaxial loading can show additional hardening
under non-proportional loading. As a result, stresses will be higher under these
loads. Models 5 and 6 predict increased fatigue damage because the increased
cyclic hardening is reflected in the maximum stress term. Models 3A and 4 do
not include any effect of the stresses from the additional cyclic hardening.
Aluminium alloys do not show the additional cyclic hardening under non-
proportional loading **(15)**. This allows a discriminating experiment to be proposed. A material such as 304 stainless steel that shows a large amount of
cross hardening would be expected to have higher fatigue damage in non-
proportional loading than in proportional loading. Aluminium alloys should
not show a large difference in fatigue lives between proportional and non-

proportional loading when compared on the basis of the maximum strain amplitude.

Variable amplitude multiaxial loading is essentially an issue of how cycles will be identified and damage computed for a complex loading history. Stress strain behaviour can be modelled for variable amplitude loading with existing non-proportional cyclic plasticity models. Hayhurst, *et al.* **(16)** and McDowell, *et al.* **(17)** have conducted experiments with alternating tension and torsion loading. Tests were conducted on 304 stainless steel. A block of loading consists of forty cycles of axial strain that are followed by forty cycles of shear strain of the same equivalent strain amplitude; the sequence is repeated until failure. Failure occurred after eighty loading blocks (320 tension cycles and 320 torsion cycles). Two distinct crack systems were observed, one for the axial strains and one for the shear strains. The fatigue life for this combined test was the same as that for an axial strain test – no interaction between the damage systems was observed. This suggests that each plane accumulates damage independently from the others. This leads to the conclusion that damage could be tracked on each plane and the interaction between damage on different planes can be neglected because neither damage system becomes dominant until late in the life.

Two level tension and torsion tests has been conducted by Robillard and Cailletaud **(18)**. Torsion cycles followed by tension cycling to failure followed a linear damage summation of tension and torsion cycles in contrast to the previous experiments where the tension and torsion cycles were summed independently. Shear cracks nucleate and grow on planes parallel and perpendicular to the specimen axis during the torsion cycles. Cracks perpendicular to the specimen axis can then grow in mode I when the tension cycles are applied, resulting in considerable interaction between the two damage systems. Tension cycles followed by torsion cycling resulted in damage summations greater than one, indicating little interaction between the tension and torsion damage systems. Here, the cracks which nucleated on 45 degree planes during the tension cycling did not propagate during the subsequent torsion loading. Damage should be tracked on each potential failure plane once a dominant crack system is established. These experiments clearly demonstrate that the interaction and growth of the damage systems must be tracked on each potential failure plane. Few models exist for tracking the growth of damage. Both tensile and shear damage could be tracked on each plane using existing damage models; however, sequencing effects exist here just as they do in uniaxial loading.

Summary

It is impossible to conduct enough experiments to validate any damage model for all loading situations. Critical experiments have been suggested for comparing damage models on the basis of how cracks nucleate and grow. The following types of models are suggested:

HCF

$$\text{case A } \tau_a(1 + \sigma_{n, \text{max}}/2\sigma_u) = f(N_{f, A})$$

$$\text{case B } \tau_a(1 + \sigma_{n, \text{max}}/2\sigma_u) = g(N_{f, B}) \tag{2}$$

LCF (shear dominated failure)

$$\gamma_a(1 + \sigma_{n, \text{max}}/\sigma_y) = f(N_f) \tag{5}$$

LCF (tensile dominated failure)

$$\sigma_{\text{max}} \varepsilon_a = f(N_f) \tag{6}$$

Common features of these models include linking damage to the physical observations of crack nucleation and growth. They are formulated so that damage is only computed when there is a cyclic strain, and they include a maximum stress term to account for mean stresses and cyclic non-proportional hardening.

References

(1) EWING, J. A. and HUMFREY, J. C. W. (1903) The fracture of metals under repeated alternations of stress, *Phil. Trans. Roy. Soc.*, **210**, 241–253.

(2) BROWN, M. W. and MILLER, K. J. (1973) A theory for fatigue failure under multiaxial stress strain conditions, *Proc. Inst. mech. Engrs*, **187**, 745–755.

(3) FINDLEY, W. N., COLEMAN, J. J., and HANDLEY, B. C. (1956) Theory for combined bending and torsion fatigue with data for 4340 steel, *Proceedings of the International Conference on Fatigue of Metals*, The Institution of Mechanical Engineers, London, pp. 150–157.

(4) FINDLEY, W. N. (1959) A theory for the effect of mean stress on fatigue of metals under combined torsion and axial load or bending, *J. of Engng Ind.*, 301–306.

(5) McDIARMID, D. L. (1972) *Failure criteria and cumulative damage in fatigue under multiaxial stress conditions*, PhD Thesis, City University, London.

(6) GOUGH, H. J., POLLARD, H. V., and CLENSHAW, W. J. (1951) Some experiments on the resistance of metals under combined stress, *Ministry of Supply, Aeronautical Research Council Reports and Memoranda No. 2522*, HMSO, London.

(7) McDIARMID, D. L. (1991) A general criterion for high cycle multiaxial fatigue failure, *Fatigue and Fracture Engng Mater.*, Vol 14, No. 4, 429–454.

(8) KANDIL, F. A., BROWN, M. W., and MILLER, K. J. (1982) *Biaxial low cycle fatigue of 316 stainless steel at elevated temperatures*, Vol 280, The Metals Society, pp. 203–210.

(9) SOCIE, D. F. and SHIELD, T. W. (1984) Mean Stress Effects in Biaxial Fatigue of Inconel 718 *J. Engng Mater. Technol.*, **106**, 227–232.

(10) FATEMI, A. and SOCIE, D. F. (1988) A critical plane approach to multiaxial fatigue damage including out-of-phase loading, *Fatigue Fracture Engng Mater. Structures* **11**, 149–165.

(11) MORROW, D. L. (1989) *Biaxial fatigue of inconel 718*, PhD. Thesis, University of Illinois at Urbana-Champaign, Illinois, USA.

(12) SMITH, R. N., WATSON, P., and TOPPER, T. H. (1970) A stress–strain parameter for the fatigue of metals *J. Mater.*, **5**, 767–778.

(13) SOCIE, D. F. (1987) Multiaxial fatigue damage models, *J. Engng Mater. Technol.*, **109**, 293–298.

(14) SOCIE, D. F., KURATH, P., and KOCH, J. L. (1989) A multiaxial fatigue damage parameter, Biaxial and multiaxial fatigue EGF3, (Edited by M. W. Brown and K. J. Miller), Mechanical Engineering Publications, London, pp. 535–550.

(15) DOONG, S. H., SOCIE, D. F., and ROBERTSON, I. M. (1990) Dislocation substructures and nonproportional hardening, *J. Engng Mater. Technol.*, **112**, 456–464.

(16) HAYHURST, D. R., LECKIE, F. A., and McDOWELL, D. L. (1985) Damage growth under nonproportional loading, *ASTM STP 853*, (ASTM, Philadelphia), pp. 688–699.

(17) McDOWELL, D. L., STAHL, D. R., STOCK, S. R., and ANTOLOVICH, S. D. (1988) Biaxial path dependence of deformation substructure of type 304 stainless steel, *Meta. Trans. A*, 1277–1293.

(18) ROBILLARD, M. and CAILLETAUD, G. (1991) Directionally defined damage in multiaxial low cycle fatigue: experimental evidence and tentative modelling, *Fatigue under biaxial and multiaxial loading*, ESIS 10, (Edited by K. F. Kussmaul, D. L. McDiarmid, and D. F. Socie), Mechanical Engineering Publications, London, pp. 103–130.

*D. L. McDiarmid**

Multiaxial Fatigue Life Prediction Using a Shear Stress Based Critical Plane Failure Criterion

REFERENCE McDiarmid, D. L., **Multiaxial fatigue life prediction using a shear stress based critical plane failure criterion,** *Fatigue Design*, ESIS 16 (Edited by J. Solin, G. Marquis, A. Siljander, and S. Sipilä) 1993, Mechanical Engineering Publications, London, pp. 213–220.

ABSTRACT A shear stress based critical plane criterion of failure, modified for the effect of normal stress on the plane of maximum range of shear stress, is used to predict fatigue lives of thin wall tubular specimens subjected to biaxial in-phase and 90 degree out-of-phase tension-torsion loading. The experimental data used is part of an extensive multiaxial fatigue programme co-ordinated by the Society of Automotive Engineers (USA).

The criterion is found to correlate the experimental data with predictions being within a factor of two. Of the forty-two experimental cases, 70 percent are predicted within a factor of 1.5.

Notation

ε_a	Axial strain amplitude
γ_a	Shear strain amplitude
σ_a	Axial stress amplitude
σ_n	Normal stress amplitude on the plane of maximum range of shear stress
$\sigma_{n, max}$	Maximum normal stress on the plane of maximum range of shear stress
τ_a	Shear stress amplitude
τ_{max}	Maximum shear stress
τ_f'	Fatigue strength coefficient
λ_ε	Strain ratio, γ_a/ε_a
λ_σ	Stress ratio, τ_a/σ_a
b	Fatigue strength exponent
k	Material constant, $t_{A, B}/2 R_m$
t_A	Reverse shear fatigue strength for case A crack growth
t_B	Reverse shear fatigue strength for case B crack growth
N_f	Cycles to failure
R_m	Ultimate tensile strength

Introduction

Many engineering components and structures in service are subjected to multiaxial fatigue stress conditions. In most of these cases uniaxial fatigue test

* Department of Mechanical Engineering, City University, Northampton Square, London, EC1V 0HB, UK.

data are used for design purposes. This is due to the expense and complexity involved in multiaxial fatigue research, both experimental and theoretical.

Much effort has been expended over many years searching for a multiaxial fatigue failure criterion which can be applied to any general multiaxial fatigue stress condition. Criteria have been found to give good correlation with particular limited multiaxial fatigue stress conditions. As yet no general criterion of failure under multiaxial stress has been found which can cope with effects such as rotating principal stress axes, non-proportional loading, and mean stress over the complete life spectrum.

Most of the multiaxial fatigue theories presented so far can be categorized as stress-based, strain based, or critical plane based. The stress and strain based criteria are essentially the classical static failure theories extended to fatigue. Better correlation of multiaxial fatigue data has been obtained using critical plane approaches. According to these theories cracks initiate and grow in particular planes and any normal stresses or strains on these planes can assist in the fatigue crack growth process. These theories have the advantage of a physical interpretation of the fatigue damage process. They can also account for mean stress effects, unlike the classical shear stress and strain theories, through the normal stress or strain term.

A modified shear stress based critical plane criterion of failure is used in the analysis presented in this paper.

Modified shear stress based critical plane criterion of failure

A general criterion for high cycle multiaxial fatigue failure, from McDiarmid (1) has been found to give good correlation with an extensive range of multiaxial fatigue experimental data from the literature under a wide variety of multiaxial loading conditions.

$$\tau_{max}/t_{A, B} + \sigma_{n, max}/2R_m = 1 \tag{1}$$

where

τ_{max} = shear stress amplitude on the plane of maximum range of shear stress (critical plane)

$\sigma_{n, max}$ = maximum normal stress on the plane of maximum range of shear stress

$t_{A, B}$ = reverse shear fatigue strength for case A, or case B crack growth, whichever is applicable

R_m = ultimate tensile strength

This criterion can be expressed as

$$\tau_{max} + k\sigma_{n, max} = \text{constant} \tag{2}$$

where k is a material constant = $t_{A, B}/2R_m$.

Life predictions

Equation (2) can be combined with Basquin's equation for uniaxial loading

$$\tau_{max} = \tau'_f(2N_f)^b \tag{3}$$

to predict fatigue lives under multiaxial loading.

The experimental data used to test the validity of the failure criterion is from Leese and Socie (2), where the fatigue behaviour of thin-walled tubular specimens subjected to biaxial-in-phase and 90 degree out-of-phase tension–torsion constant amplitude loading in strain control at room temperature using 1045 HR steel, was investigated. These tests were part of an extensive biaxial fatigue testing programme organized by the Society of Automotive Engineers (USA).

The cyclic constants used were those obtained from thin-walled tube specimens under axial cyclic loading ($\lambda_\varepsilon = 0$ in Table 1). These values are: $\tau'_f = 627.7$ MN/m^2 and $b = -0.111$. Substitution in equation (3) gives

$$\tau_{max} + k\sigma_{n,\,max} = 627.7(2N_f)^{-0.111} \tag{4}$$

and hence

$$N_f = \{581.2/(\tau_{max} + k\sigma_{n,\,max})\}^{9.01} \tag{5}$$

Biaxial in-phase fatigue test results from (2) and predicted lives using equation (5) are shown in Table 1.

Biaxial 90 degree out-of-phase fatigue test results for both sinusoidal and trapezoidal strain histories from (2) and predicted lives using equation (5) are shown in Table 2. Examination of Table 2 in (2) shows that for $\lambda_\varepsilon = 0.5$ and 1 the greatest values of $(\tau_{max} + k\sigma_{n,\,max})$ occur at the instant when the axial stress is a maximum and the torsion stress is zero. It is noted that the maximum shear stress produced when torsion is a maximum and axial load is zero, is less than the maximum shear stress produced when axial stress is a maximum and torsion is zero. When $\lambda_\varepsilon = 2$ and $\lambda_\sigma > 0.5$ the maximum value of $(\tau + k\sigma_n)$ taking k for the relevant life can be found as follows. For out-of-phase angle of 90 degrees

$$\tau + k\sigma_n = \tau \sin(\omega t + 90) + k\sigma \sin \omega t.$$

$$= \tau \cos \omega t + k\sigma \sin \omega t$$

For a maximum value of $(\tau + k\sigma_n)$, then

$$\frac{d(\tau + k\sigma_n)}{d(\omega t)} = 0 = -\tau \sin \omega t + k\sigma \cos \omega t$$

i.e., when $\tan \omega t = k\sigma/\tau = k/\lambda_\sigma$.

Hence maximum values of $(\tau + k\sigma_n)$ can be found for different values of $\lambda_\sigma = \tau_a/\sigma_a$.

Table 1 Biaxial in-phase fatigue test results, from (2)

Spec. No.	λ_ε	λ_σ	σ_a MN/m²	τ_a MN/m²	σ_n MN/m²	τ_{MAX} MN/m²	$t_A/2R_m = k$	$(\tau_{MAX} + k\sigma_n)$ MN/m²	$N_{f,TEST}$	$N_{f,PRED}$	$\dfrac{N_{f,PRED}}{N_{f,TEST}}$
4553	0.0	0.00	459	0	230	230	0.2	276.0	1108	820	0.74
4527	0.0	0.00	448	0	224	224	0.2	268.8	1137	1041	0.92
4505	0.0	0.00	384	0	192	192	0.18	226.6	4600	4850	1.05
4507	0.0	0.00	377	0	188.5	188.5	0.18	232.4	4959	3862	0.78
4545	0.0	0.00	354	0	177	177	0.16	205.3	8680	11802	1.36
4552	0.0	0.00	279	0	139.5	139.5	0.13	156.2	78271	138520	1.77
4511	0.0	0.00	274	0	137	137	0.13	154.8	142541	150223	1.05
4547	0.0	0.00	240	0	120	120	0.13	135.6	$>2 \times 10^6$	495352	—
4524	0.5	0.20	430	84	215	230.8	0.2	277.0	1259	794	0.63
4523	0.5	0.18	340	60	170	180.3	0.16	207.5	11777	10721	0.91
4528	0.5	0.20	261	53	130.5	140.9	0.13	157.9	92000	125646	1.36
4516	0.5	0.22	265	57	132.5	144.2	0.13	161.4	115500	103125	0.89
4519	0.5	0.19	244	47	112	121.4	0.13	136.0	611800	482379	0.79
4525	1.0	0.36	383	138	191.5	236.0	0.2	274.2	1616	870	0.54
4520	1.0	0.36	304	109	152	187.0	0.16	211.3	10380	9104	0.88
4515	1.0	0.37	308	113	154	191.0	0.16	215.6	11610	7593	0.65
4550	1.0	0.40	235	95	117.5	151.1	0.13	163.6	90000	91282	1.01
4514	1	0.39	238	93	119.0	151.0	0.13	166.5	123500	77917	0.63
4554	1	0.39	202	79	101.0	128.2	0.13	141.3	393600	341816	0.87
4517	1	0.40	212	85	106.0	135.9	0.13	149.7	595600	203154	0.34
4526	2	0.70	289	202	144.5	248.4	0.20	277.3	1758	786	0.45
4503	2	0.68	234	159	117.0	197.4	0.16	216.1	16890	7436	0.44
4501	2	0.68	232	158	116.0	196.0	0.16	214.5	20030	7951	0.40
4522	2	0.73	179	131	89.5	158.7	0.13	170.3	98750	63581	0.64
4548	2	0.73	179	132	89.5	159.5	0.13	171.1	87500	60952	0.70
4521	2	0.79	147	116	73.5	137.3	0.13	146.9	545800	240830	0.44
4509	4	1.28	139	179	69.5	192.0	0.16	203.1	19769	13005	0.66
4530	10	3.67	45	165	22.5	166.5	0.13	169.4	66808	66690	1.00
4549	∞	∞	0	255	0.0	255.0	0.20	255.0	890	1674	1.88
4506	∞	∞	0	201	0.0	201.0	0.16	201.0	8710	14282	1.64
4551	∞	∞	0	169	0.0	169.0	0.13	169.0	57369	68126	1.19
4512	∞	∞	0	171	0.0	171.0	0.13	171.0	102083	61274	0.60
4518	∞	∞	0	152	0.0	152.0	0.13	152.0	1.01×10^6	177077	0.18

Table 2 90 degree out-of-phase fatigue test results, from (2)

Spec. No.	λ_ε	λ_σ	σ_a MN/m²	τ_a MN/m²	σ_n MN/m²	τ_{MAX} MN/m²	$t_A/2R_m = k$	$(\tau_{MAX} + k\sigma_n)$ MN/m²	$N_{f,\,TEST}$	$N_{f,\,PRED}$	$\dfrac{N_{f,\,PRED}}{N_{f,\,TEST}}$
45A3	0.5	0.41	364	182	182	182.0	0.18	214.2	5 260	8 052	1.53
4583	0.5	0.33	281	140.5	140.5	140.5	0.14	160.1	58 525	110 920	1.90
4580	1.0	0.51	377	188.5	188.5	188.5	0.18	222.4	5 119	5 740	1.12
45B2	1.0	0.51	289	144.5	144.5	144.5	0.14	164.7	49 143	85 934	1.75
4586	1.0	0.51	285	142.5	142.5	142.5	0.14	162.5	64 652	97 003	1.50
45A5	1.0	0.46	232	116	116	116.0	0.13	131.1	1.39×10^6	671 388	0.48
45D4	2.0	0.59	345	204	100.7	194.6	0.18	212.9	5 262	8 506	1.61
45D1	2.0	0.67	250	167	51.3	164.0	0.14	171.3	34 718	60 314	1.73
45B5	2.0	0.61	272	166	56.6	162.4	0.13	169.7	38 925	65 636	1.69
4588	2.0	0.69	190	131	35.2	128.8	0.13	133.3	613 554	577 902	0.94
Trapezoidal strain history											
4584	2	0.58	380	220	112.1	210.5	0.18	230.7	4 350	4 126	0.95
45D2	2	0.65	274	177	65.5	172.9	0.16	183.3	18 325	32 770	1.79
45A4	2	0.72	207	148	39.5	146.3	0.14	151.9	91 455	178 130	1.95

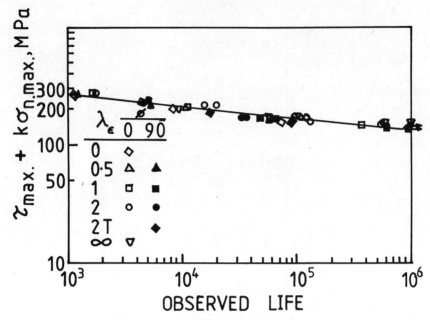

Fig 1   Fatigue life correlation with modified shear stress failure criterion

The degree of correlation of this failure criterion can be seen to be good in Fig. 1. Also as shown in Fig. 2 good correlation is obtained with life predictions being mainly conservative up to a factor of two for the in-phase loading and non-conservative up to a factor of two for the out-of-phase loading.

Discussion

The observed life versus predicted life plots are justified up to an observed life of about 5×10^5 cycles since the uniaxial cyclic parameters (τ_f', b) used in constructing these plots are obtained from uniaxial strain–life plots valid only in the low cycle fatigue life region. In fact there are few data points at lives $> 5 \times 10^5$ cycles.

It is seen from Fig. 2 that the torsion ($\lambda_\varepsilon = \infty$) test data gives the most non-conservative life predictions. This could be influenced by the failure definition as well as the anisotropy resulting from inclusions which act as initiation sites for longitudinal shear cracks. The possible influence of the different cracking modes existing under uniaxial and pure torsional loading conditions for the material tested is discussed in (2).

From Fig. 1 it can be seen that out-of-phase loading tends to be more damaging than in phase loading and hence life predictions for out-of-phase loading are non-conservative. In out-of-phase loading all planes are subjected to near maximum shear stress conditions and thus damage is likely to be more severe than for in-phase loading. The out-of-phase loading case is further compli-

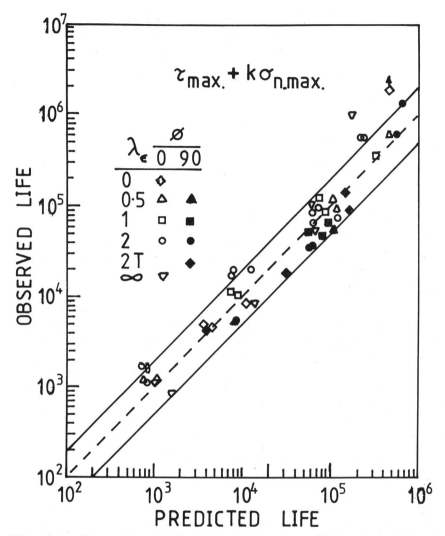

Fig 2 Fatigue life correlation with life predictions based on the modified shear stress failure criterion

cated by the fact that maximum normal stresses do not occur simultaneously with maximum shear stresses in any plane. However, the analysis used maximizes the $(\tau + k\sigma_n)$ critical parameter for each particular loading condition.

The value of $k = t_A/2R_m$ used in each case is that for t_A at the particular observed life. For all the combined axial and torsion loading conditions tested type A cracks growing along the surface occur. Values of $t_A/2R_m$ range from 0.2 at 10^3 cycles to 0.13 at 10^5 cycles. In fact if $t_A/2R_m = 0.13$ is used in all cases, better correlation is obtained over the 10^3 to 10^5 cycle life range, that is in-phase life predictions are less conservative.

Table 3　Correlation of experimental data from (2) with criteria of failure

Criterion of failure	In phase	Out-of phase	Non-conservative　←|→　Conservative	
Maximum shear stress	✓			├———→ 7
		✓	2 ←→ ├→ ├———→ 5	
Octahedral shear stress	✓			├→ 2
		✓	2 ←→ ├→ 2	
Maximum shear strain	✓			├———→ 3
		✓	3.5 ←———┤	
Octahedral shear strain	✓			├→ 2
		✓	3.5 ←———┤	
Zamrick (4) total strain	✓			├→ 2
		✓	4.5 ←———┤	
Brown and Miller (3)	✓			├→ 2
		✓	2 ←┤	
McDiarmid (1) modified shear stress	✓			├→ 2
		✓	2 ←┤	

A comparison of the degree of correlation found using a number of failure criterion from (2) is shown in Table 3. The octahedral shear stress criterion approach, (Brown and Miller (3)), and the modified shear stress criterion presented here give the best correlations, all within conservative/non-conservative factors of two. The octahedral shear stress criterion has the disadvantage of not being able to account for the effect of mean stress, whereas the other two criteria can account for mean strain/stress within their normal strain/stress terms. From a study of (2) it would appear that the formulation and use of the modified shear stress criterion proposed here is more straightforward than the Brown and Miller approach.

Conclusion

The modified shear stress based critical plane criterion of failure

$$\tau_{\text{max}} + k\sigma_{\text{n, max}} = \text{constant}$$

is reasonably accurate in correlating both the in-phase and 90 degree out-of-phase tension-torsion SAE experimental data. Life predictions using this approach were mainly conservative up to a factor of two for in-phase loading and non-conservative up to a factor of two for 90 degree out-of-phase loading.

References

(1) McDIARMID, D. L. (1991) A general criterion of high cycle multiaxial fatigue failure, *Fatigue Fracture of Engng Maters Structures*, **14**, 429–453.
(2) LEESE, G. E. and SOCIE, D. (1989) *Multiaxial fatigue: analysis and experiments*, Society of Automotive Engineers, USA, pp. 121–137.
(3) BROWN, M. W. and MILLER, K. J. (1973) A theory for fatigue under multiaxial stress-strain conditions, *Proc Instn mechanical Engrs*, **187**, 745–755.
(4) ZAMRIK, S. Y. and FRISHMUTH, R. E. (1973) The effect of out-of-phase biaxial strain cycling on low cycle fatigue, *Expl Mech.*, **13**, 204–208.

J. Andersons*, V.Limonov†, and V. Tamužs*

Effect of Phase Lag on the Cyclic Durability of Laminated Composites

REFERENCE Andersons, J., Limonov, V., and Tamužs, V., **Effect of phase lag on the cyclic durability of laminated composites,** *Fatigue Design,* ESIS 16 (Edited by J. Solin, G. Marquis, A. Siljander, and S. Sipilä) 1993, Mechanical Engineering Publications, London, pp. 221–229.

ABSTRACT Theoretical and experimental results on fatigue of laminated fibre-reinforced composites under out-of-phase biaxial cyclic loading are presented. Experiments were carried out on tubular filament-wound samples of an epoxy matrix/organic (i.e., Kevlar-type) fibre composite. Fatigue strength under two different loading modes, namely cyclic torsion combined with axial tension or compression, was investigated for phase lags $\psi = 0$, $\pi/2$, π. Durability was shown to decrease with increasing phase shift both for axial tension ($R = 0.1$) and compression ($R = 10$).

Matrix failure criterion was proposed for unidirectionally reinforced ply and the ply discount method was modified to account for phase lag. Calculated S–N curves agreed reasonably well with experimental data.

Introduction

Composite materials are widely used as primary load bearing structural elements in different technical applications. This necessitates the investigation of the effect of various design and service factors on the durability of composites. Thus, in references **(6)** and **(1)** the effect of reinforcement angles and the stacking sequence of undirectionally reinforced plies on the fatigue strength of laminate were studied experimentally. In references **(13)(11)(12)(7)** and **(1)** the influence of a number of service factors (loading frequency, temperature and moisture content, the stress ratio of cyclic loading, mutliaxiality of proportional cyclic loading) on the fatigue life of uni- and multidirectionally reinforced composites was investigated.

The diversity of laminated fibre-reinforced composites together with pecularities of their applications, make it impossible to assess their load bearing capacity solely by means of experiments. Even for only one material system, such an experimental investigation would be very time and material consuming. Therefore, the task of summarizing the experimental data obtained and developing a reliable theoretical framework for fatigue life prediction becomes acute. As the first step in this direction, implementation of the ply discount method seems to be promising **(1)(2)(3)(7)**.

Little effort has been made so far to study the effect of complex loading (most of the fatigue research is confined to the proportional cyclic loading mode, e.g. **(1)**), while actual multiaxial service loads frequently lead to a complex stress state with non-proportional cyclic stress tensor components. The simplest of the regular complex loading modes is probably the out-of-

* Institute of Polymer Mechanics, Riga, Latvia.
† CNIISM, Hotkovo, Russia.

phase mode. A remarkable influence of the phase lag on the rate of fatigue damage accumulation in composite laminates is observed (10). The primary objectives of this paper are the experimental investigation of the phase lag effect on cyclic durability and the verification of the ply discount model for loading modes considered.

Failure criterion

The assessment of the fatigue life of an isotropic homogeneous material becomes much more sophisticated under a complex loading path (9). The highly organized structure of laminated fibre-reinforced composites permits a simplification of the problem of lifetime prediction under a plane stress state. It is necessary to determine only the fatigue life of structural elements – unidirectionally reinforced plies having two distinct failure modes, i.e. matrix cracking and fibre fracture. Fatigue strength of the laminate can then be determined by a ply-by-ply failure analysis (1).

The fatigue failure criterion of the unidirectionally reinforced ply is derived using the so-called geometrical approach of phenomenological fracture mechanics (8). The critical state of stress under quasi-static loading can be represented by a surface $F(\sigma_{ij}/\Sigma) = 1$ in the stress space σ_{ij}; Σ stands for experimentally obtained values of strength under a definite set of simple loading modes in terms of which the parameters of the failure surface equation are expressed. It is assumed that, regardless of the loading path, failure takes place when the stress trajectory (i.e., the trajectory of the point denoting current stress state in the stress space) touches the failure surface.

Cyclic loading implies a closed stress path which does not intersect with the static failure surface. This causes the accumulation of microdefects leading to a gradual deterioration of elastic and strength characteristics, i.e., fatigue degradation of the material. Residual strength theories treat the degradation of the load bearing capacity of the material undergoing cycling as a reduction of residual strength, that is the remaining static strength after fatigue loading of some duration. When residual strength coincides with the maximum value of the cyclic load, fatigue failure takes place, i.e., the last cycle before failure may be considered as the static strength test. Following this concept, the equation for the residual strength surface can be obtained from that for static strength, substituting static strengths Σ_s by corresponding residual strengths Σ_r (thus we come to a failure surface shrinking gradually during fatigue loading). Taking into account the relative scarceness of available experimental information on the residual strength degradation of unidirectionally reinforced composites, the failure criterion should be written in terms of more convenient S–N curves. Generally $\Sigma_s > \Sigma_r(N) > \Sigma(N)$ ($\Sigma(N)$ denotes ordinary fatigue diagram), see Fig. 1. Therefore, fatigue strength can serve as a conservative estimate of the residual strength for the same number of elapsed cycles. Thus, the fatigue failure criterion finally takes the form $F(\sigma_{ij}/\Sigma_{ij}(N)) = 1$. This expres-

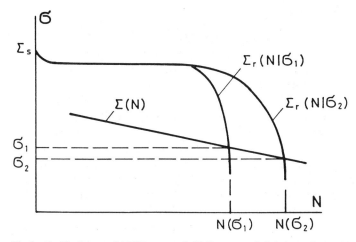

Fig 1   Residual strength $\Sigma_r(N)$ curves, S–N diagram and their interrelationship

sion agrees with those proposed earlier **(5)(15)(16)** for the case of proportional loading (and obtained by substituting the static strengths with fatigue strengths in the static failure criteria), but differs from them in that it can be applied to the case of complex loading path as well.

Let us apply the expression obtained to out-of-phase cyclic loading, using Hashin's failure criterion **(5)**. The criterion of fibre fracture does not change form due to its linearity, whereas matrix cracking criterion takes the following form

$$\left(\frac{\sigma_{2m} + \sigma_{2a} \sin{(\omega t)}}{\Sigma_2(N)}\right)^2 + \left(\frac{\sigma_{12m} + \sigma_{12a} \sin{(\omega t + \psi)}}{\Sigma_{12}(N)}\right)^2 = 1 \tag{1}$$

where: Σ_2 and Σ_{12} stand for fatigue strength under the loading with stress ratios $(\sigma_{2m} - \sigma_{2a})/(\sigma_{2m} + \sigma_{2a})$ and $(\sigma_{12m} - \sigma_{12a})/(\sigma_{12m} + \sigma_{12a})$, respectively; t is the duration of loading; and $N = t*f$ is the number of cycles. For proportional cycling, equation (1) can easily be transformed into an expression coinciding with the one proposed in **(5)**. Verification of the latter is carried out for the unidirectional glass and organic (Kevlar-type) fibre reinforced plastics **(5)(14)**. Due to the lack of experimental information concerning the out-of-phase fatigue of unidirectional composites, the validity of the more general criterion (1) can be evaluated only indirectly by comparing predicted S–N curves of laminated composites with the test results.

It should be mentioned that the implementation of the approach outlined above for fatigue life prediction at even simpler complex loading mode, (combination of static shear and cyclic compression), yielded good agreement with the theory and test results of fibre glass composites **(3)**.

Materials and methods

Tubular thin-wall specimens of an organic (Kevlar-type) fibre/epoxy matrix composite with layup (± 30 degrees; 90 degrees; 90 degrees; ± 30 degrees) were tested. The geometrical characteristics and the technology of manufacture of the specimens are described in detail in (5). Fibre volume fraction of the helically wound plies (reinforcement direction $\varphi = 30$ degrees) was 60–65 percent, that of the hoop plies ($\varphi = 90$ degrees) was 65–70 percent.

Experiments were carried out by a universal test machine Instron-1343, designed for tensile, compressive, and torsional loading with phase lag range $\psi = 0, \ldots \pi$. Uni- and biaxial fatigue tets were performed under load control at fixed frequency of $f = 17$ Hz; the stress ratio for axial tension and torsion was $R = 0.1$, and compression $- R = 10$. Tests were terminated at $N = 10^6$ cycles if failure had not occurred.

Laminated tubular samples were considered to have failed when the displacement amplitude of the active grip of the test setup exceeded that recorded during the first loading cycle by 15 percent. The percentage is arbitrarily chosen. This is not expected to lead to any major errors in the assessment of the lifetime which is usually defined as the number of cycles necessary to cause separation of the sample in two parts and thus ultimate failure, since the last stage of the active fatigue failure process was found to last only for 500–600 cycles. During that stage, an abrupt increase in temperature and deformation took place, leading to some decrease of the preset load level as the feedback circuit of the test setup failed to cope with it.

Results and discussion

Tests under uniaxial loading were carried out to evaluate the static and fatigue strength of the composite under consideration. The testing procedure is described in (6). Axial tension only, compression only, and torsion only were applied; corresponding S–N curves are presented in Fig. 2. Failure modes and mechanism of a similar composite of layup (± 30 degrees; 90 degrees; 90 degrees; ± 30 degrees) under the loading modes mentioned above are described in (6).

In order to ensure a comparable influence of both axial and torsional load components on the fatigue damage accumulation under biaxial loading, the ratio of the loads was chosen as follows. The lifetime equal for all the uniaxial loading modes was chosen to be $N^* = 30\,000$ and the corresponding fatigue strengths for axial tension $\sigma^{*(+)}_{x,\,max}$, compression $\sigma^{*(-)}_{xx,\,min}$ and torsion $\sigma^{*}_{12,\,max}$ were determined from fatigue diagrams (Fig. 2). Basic loading beams for proportional loading (shown schematically in Fig. 3) were then determined through the ratios of these strengths: for simultaneous tension and torsion $\alpha_1 = \text{arctg}(\sigma^{*(+)}_{xx,\,max}/\sigma^{*}_{xy,\,max})$, for compression and torsion $\alpha_2 = \text{arctg}(\sigma^{*(-)}_{xx,\,min}/\sigma^{*}_{xy,\,max})$.

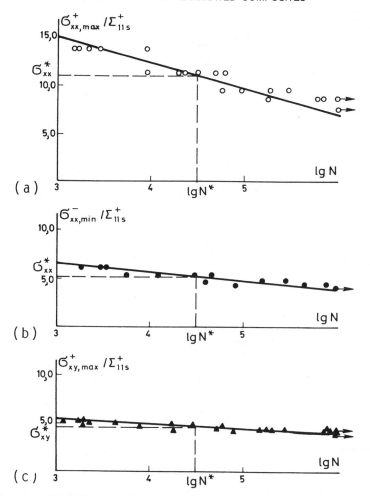

Fig 2 Experimental S–N diagrams of laminated composite of lay-up (± 30 degrees; 90 degrees; 90 degrees; ± 30 degrees) under: (a) uniaxial cyclic tension ($\bigcirc$); (b) compression ($\bullet$); (c) shear ($\blacktriangle$); ———— regression curves

Biaxial loading of the composite specimens was performed under in-phase, out-of-phase (phase lag $\psi = \pi/2$) and antiphase modes. Load paths are shown in Fig. 3. Both in-phase and anti-phase loading lead to linear load trajectories (either proportional or non-proportional depending on the value of the loading beam parameter α), while in the case of $\psi = \pi/2$ the load path is an ellipse. In all cases the specimens failed in a shear instability mode preceded by localized splitting of helically wound layers and limited fibre bundle pullout under tension, localized debonding and bending of some fibre bundles of the helical plies under compression. The fatigue data obtained are shown in Fig. 4, together with the fatigue life diagrams predicted theoretically. The data

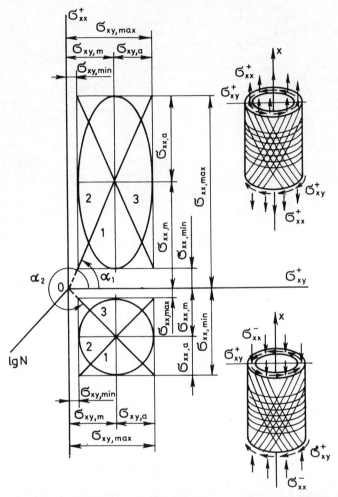

Fig 3 Directions of axes and loading modes of composite laminate under combined axial and torsional loading: (1) in-phase loading; phase lag $\psi = 0$; (2) out-of-phase loading; $\psi = \pi/2$; (3) loading in anti-phase; $\psi = \pi$

suggest that the fatigue strength of the composite under consideration decreases with increasing phase shift. The reduction may constitute as much as 50 percent of the strength under in-phase loading for axial compression (Fig. 4(b)), and the experimental curves seem to become steeper at higher values of phase lag.

Theoretical curves shown in Fig. 4 and 5 were calculated according to **(1)** using the modulae and Goodman diagrams of the unidirectionally reinforced composite at uniaxial loading in the fibre direction, transverse to them and inplane shear **(7)**. The latter experimental information is required since the stress ratio in the plies depends not only on that of the cyclic loads but on the

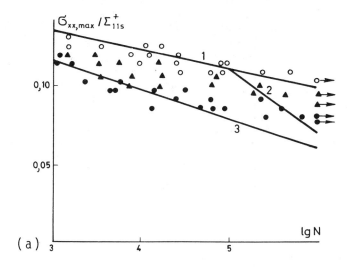

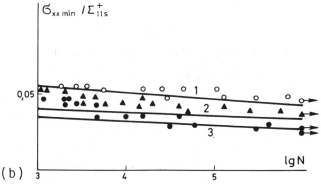

Fig 4　S–N diagrams of laminated composite of lay-up (± 30 degrees; 90 degrees; 90 degrees; ± 30 degrees) under: (a) combined axial tension and shear for loading beam with $\alpha = 67$ degrees; (b) axial compression and shear for $\alpha = 312$ degrees. Phase lag $\psi = 0$ (○); $\pi/2$ (▲); π (●). ———— theoretical predictions

phase lag between them as well. The available experimental information (7) concerns a composite with fibre volume fraction μ equal to that of the helical plies. Because transverse-to-fibre fatigue characteristics of unidirectional Kevlar-type composite are virtually independent on μ (at least for the range of μ of interest in this case – compare corresponding S–N curves for $\mu = 60$–65 percent (6) and $\mu = 65$–70 percent (14)) due to the closeness of the values of radial modulus of the fibre and modulus of the matrix, only longitudinal characteristics should be changed to accommodate them for the hoop plies possessing higher μ. This was done by means of the 'rule of mixture', shown to be valid for longitudinal fatigue strength of unidirectional composite (4).

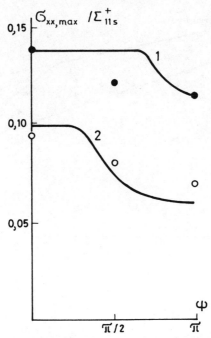

Fig 5   Variation of fatigue strength of composite laminate of lay-up (± 30 degrees; 90 degrees; 90 degrees; ± 30 degrees) with phase lag between axial and shear loads for loading beam $\alpha = 67$ degrees and fixed endurance of $10^3 - 1$ ($\bullet$) and $10^6 - 2$ ($\bigcirc$) cycles

Calculations suggest that under combined torsion and compression the failure of the laminate is associated with the fibre fracture of a pair of helical plies. The effect of phase lag in this case is manifested through the variation of the stress ratio in the fibre direction of the critical plies. Under axial tension the theoretical failure mode depends upon the phase shift of the loads. At small values of ψ the durability of the specimen is determined by the strength of fibres, but with the increase of ψ the failure mode changes to the matrix-controlled load.

References

(1) ANDERSON, J. A., LIMONOV, V. A., and TAMUZH, V. P. (1989) Fatigue of laminated composites with various reinforcement systems. Part 2: Planar stress state and calculation model, *Mech. composite Mater.*, **25**, 442–450.

(2) ANDERSON, J. A., LIMONOV, V. A., and TAMUZH, V. P. (1990) Failure during axial loading of a plastic reinforced at oblique angles with organic fibres, *Mech. composite Mater.*, **26**, 182–188.

(3) ANDERSONS, J., MIKELSONS, M., TAMUŽS, V., and LIMONOV, V. (1993) Fatigue of laminated composites under complex cyclic loading, *Proceedings of the ninth international conference on composite materials*, Vol. 5, Woodhead Publishing Limited, pp. 763–768

(4) DHARAN, C. K. H. (1975) Fatigue failure in graphite fibre and glass fibre–polymer composites, *J. Mater. Sci.*, **10**, 1665–1677.

(5) HASHIN, Z. and ROTEM, A. (1973) A fatigue failure criterion for fiber reinforced materials, *J. composite Mater.*, **7**, 443–464.

(6) LIMONOV, V. A., PEREVOZCHIKOV, V. G., and TAMUZH, V. P. (1988) Fatigue of laminated composites with various reinforcement systems. Part 1: Experimental results, *Mech. composite Mater.*, **24**, 585–594.

(7) LIMONOV, V. A. and ANDERSON, J. A. (1991) Influence of stress ratio on fatigue strength of laminated composite, *Mech. composite Mater.*, **27**, 276–283.

(8) MALMEISTER, A., TAMUZH, V., and TETERS, G. (1977) *Mechanik der Polymerwerkstoffe*, Akademie Verlag, Berlin, p. 592.

(9) MILLER, K. J. and BROWN, M. W. (1984) Multiaxial fatigue – a brief review, Proc. ICFG. 1(1984), pp. 31–56.

(10) OHLSON, N. G. (1985) Damage in biaxial fatigue of composites, *Proceedings of the fifth international conference on composite materials*, The Metallurgical Society, pp. 31–56.

(11) OLDYREV, P. P., and MALINSKII, A. M. (1983) The effect of the temperature on multicycle fatigue of an organoplastic, *Mech. composite Mater.*, **19**, 261–266.

(12) OLDYREV, P. P. (1983) The effect of the moisture on multicycle fatigue of reinforced plastics, *Mech. composite Mater.*, **19**, 446–456.

(13) OLDYREV, P. P. and APINIS, R. P. (1983) The effect of the loading frequency on multicycle fatigue of an organoplastic, *Mech. composite Mater.*, **19**, 629–633.

(14) PEREVOZCHIKOV, V. G., LIMONOV, V. A., PROTASOV, V. D., and TAMUZH, V. P. (1988) Static and fatigue strength of unidirectional composites under the combined effect of shear stress and transverse tension–compression stresses, *Mech. composite Mater.*, **24**, 638–644.

(15) RABOTNOV, Ju. N., *et al.* (1985) Cyclic strength of unidirectional carbon-reinforced plastic in tensile loading at an angle in relation to the reinforcement direction, *Mech. composite Mater.*, **21**, 158–162.

(16) SIMS, D. F. and BROGDAN, V. H. (1977) Fatigue behavior of composites under different loading modes, *ASTM STP 636*, ASTM, Philadelphia, pp. 185–205.

A. V. Troshchenko*, K. J. Miller*, and U. S. Fernando*

A Simple Criterion for Multiaxial Fatigue Life Prediction under In-Phase and Out-of-Phase Tension–Torsion

REFERENCE Troshchenko, A. V., Miller, K. J., and Fernando, U. S., **A simple criterion for multiaxial fatigue life prediction under in-phase and out-of-phase tension–torsion,** *Fatigue Design*, ESIS 16 (Edited by J. Solin, G. Marquis, A. Siljander, and S. Sipilä) 1993, Mechanical Engineering Publications, London, pp. 231–240.

ABSTRACT A new approach is proposed for low-cycle fatigue life prediction which is based on a parameter that evaluates changes of a modified octahedral shear strain over one cycle of stabilized loading. A comparison with experimental results shows good correlation both for in-phase and out-of-phase loading.

Notation

ε_{ij}	Time-dependent strain components
ε_z, ε_θ, ε_r	Axial, circumferential, and radial strain components
$\gamma_{z\theta}$, $\gamma_{\theta r}$, γ_{rz}	Shear strains on the $z\theta$, θr, rz planes, respectively
ε_a', γ_a	Axial and shear strain amplitudes
ε_1, ε_2, ε_3,	Principal strains; $\varepsilon_1 > \varepsilon_2 > \varepsilon_3$
$\lambda = \gamma_a/\varepsilon_a$	Shear to axial strain ratio
ξ	Strain state parameter
γ_{oct}	Octahedral shear strain
$\bar{\gamma}_{oct}$	Modified octahedral shear strain
r	Normalized octahedral shear strain
N_f, N_f^*	Actual and normalized number of cycles to failure, respectively
ν^*	Poisson's ratio
ϕ	Phase lag between ε_a and γ_a in a tension–torsion test
A_i	Cyclic material properties
T	Cycle period

Superscripts

pp, t	Push–pull loading and torsion loading, respectively

Introduction

In recent years there has been a significant increase in the number of publications in the field of biaxial and multiaxial fatigue, e.g., (1)–(3), particularly with regard to fatigue life prediction under complex loading conditions. This, in part, is due to new safety requirements for structural design which may also involve a reduction in both structural and material weight plus variable temperature and environmental effects.

* SIRIUS, Faculty of Engineering, University of Sheffield, Mappin Street, Sheffield S1 3JD, UK.

These considerations lead to complex experiments which attempt to duplicate actual multiaxial loading histories using test facilities that permit a continuous changing of stress and strain states, including a re-orientation of principal stress and strain axes. Consequently, following a wide range of experiments to investigate the nature of fatigue failure under complex loading, several different parameters have been introduced to assist in the development of life prediction methodologies. These methods can be divided into two basic categories:

(1) procedures that use damage parameters which specifically relate to fatigue crack growth behaviour;
(2) procedures that are based on equivalent stress–strain parameters.

With respect to the first category it is historically recognized (4) that fatigue damage, in physical terms, can be related to crack initiation and propagation. However, information on crack growth behaviour under multiaxial loading is rare and is seldomly sufficient to develop a fatigue life prediction procedure that can be safely applied to industrial situations.

The second category invariably relates to the establishment of a relationship between number of cycles to failure and an equivalent stress (or strain) parameter which can easily be evaluated from knowledge of the applied loads and/or resulting strains. In this case it is essential to introduce a function which defines a parameter that takes into account stress–strain history and the cyclic properties of the material.

Since fatigue lifetime, N_f, reflects both the initiation and propagation behaviour, both the above methodologies can be united into a single approach if the parameter of the second method is related to the number of cycles to failure. An important requirement from the viewpoint of engineering design is that the coupled function should be able to correlate data for any complex loading path and that the life prediction can be conducted with data generated from simple tests e.g., uniaxial push–pull and/or torsion loading. Such a criteria can generally be expressed as

$$\int (\varepsilon_{ij}, N_f, A_i) = \begin{cases} \int (\varepsilon_a, N_f, A_i) - \text{for push-pull loading} \\ \int (\gamma_a, N_f, A_i) - \text{for torsion loading} \end{cases} \tag{1}$$

In this equation ε_{ij} represents time-dependent strains, N_f is number of cycles to failure, ε_a and γ_a are axial and shear strain amplitudes under reversed push-pull and torsion loading, respectively, while A_i represents the cyclic properties of the material.

In this paper an attempt is made to analyse the results obtained by different researchers (5)–(9) who used several different types of steels under a wide range of in-phase and out-of-phase loading conditions. A new equivalent strain parameter is introduced which gives good correlation with the above referenced experimental data. The parameter is based on a modified octahedral shear strain, and can be implemented from sole knowledge of endurance data for reversed tension–compression and torsion tests.

Proposed criterion and discussion

Several new equivalent strain criteria have been introduced in recent investigations (5)–(9) because other criteria like Tresca, octahedral shear strain, and Rankine, failed to predict the fatigue life, even for in-phase loading. The inadequacy of the octahedral shear strain criteria for correlating multiaxial data, particularly for the low-cycle fatigue regime, has been widely discussed (7)–(11), and this inadequacy is clearly apparent from the in-phase tension–torsion results shown in Fig. 1(a). The unsuitability of the octahedral shear strain

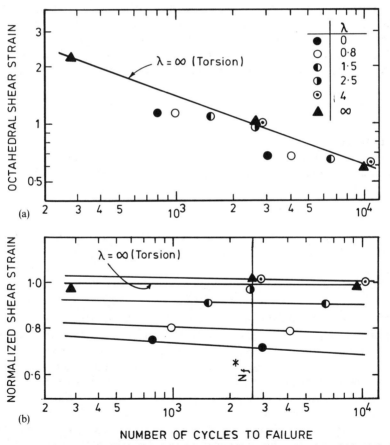

Fig 1 Tension–torsion fatigue endurance data for an En 15R steel as a function of (a) octahedral shear strain and (b) normalized shear strain

criterion can be visualized by evaluating a normalized octahedral shear strain parameter

$$r(\xi) = \frac{\gamma_{oct}}{\gamma_{oct}^t} \quad \text{where } \xi = \xi(\varepsilon_1, \varepsilon_2, \varepsilon_3) \tag{2}$$

for different multiaxial stress–strain conditions. Here ξ defines the multiaxial state of strain and ε_1, ε_2, ε_3 are the three principal strains. The terms γ_{oct} and γ_{oct}^t are, respectively, the octahedral shear strain for an arbitrary strain state and the torsion loading state corresponding to the same fatigue life. As shown by Jordan (12) for different materials, the value of r^{pp} (i.e, the value of the parameter r for push–pull loading) can deviate significantly from unity predicted by the octahedral shear strain criterion.

It can be seen from Fig. 1(b) that, for the whole range of fatigue lives, the value of the ratio r is minimum for the case of uniaxial loading and increases with the value of the strain ratio λ, where λ is the ratio of torsional (γ_a) and axial (ε_a) strain amplitudes for the tension–torsion tests. This observation was found to be valid for all the fatigue data which have been examined (5)–(9).

Therefore, let us introduce a function $\Psi(\xi, N_f)$ which describes the changes of the parameter r with strain state and satisfies the condition

$$\Psi(\xi, N_f) = \frac{1}{r(\xi)} \tag{3}$$

for all values of N_f.

The proposed equivalent strain parameter for correlating multiaxial data may then be chosen in the form of a modified octahedral shear strain, expressed as

$$\bar{\gamma}_{oct}(\varepsilon_{ij}, N_f) = \Psi(\xi, N_f)\gamma_{oct}(\varepsilon_{ij}) \tag{4}$$

Generalization of this approach for non-proportional loading leads to an integrating procedure over a complete loading cycle, i.e.

$$\bar{\gamma}_{oct}(\varepsilon_{ij}, N_f) - \int_0^T \Psi(\xi, N_f)\dot{\gamma}_{oct}(\varepsilon_{ij}) \, dt \tag{5}$$

where T is the period of the cycle, and $\dot{\gamma}_{oct}$ represents the time derivative of the parameter γ_{oct}. By integrating over a complete cycle of stabilized loading the proposed criterion takes into account time- and path-dependence of the loading history, and so can be used for both in-phase and out-of-phase loading cases. Unlike most available equivalent strain criteria this approach also takes into account changes of the function with lifetime, and so can be related to the relevant dominant physical process of fatigue damage at the different stages of life. Equation (5) represents the general form of the proposed criterion but the available experimental data only allow the function $\Psi(\xi, N_f)$ to be evaluated for certain loading conditions.

For the case of tension–torsion, when the intermediate principal strain ε_2 is normal to the surface of the specimen, it is suggested that the strain state can be defined by

$$\xi = \frac{\varepsilon_3}{\varepsilon_1} \tag{6}$$

where ε_1 and ε_3 are the maximum and minimum principal strains. The octahedral shear strain can be calculated as

$$\gamma_{oct} = \frac{2}{3} \sqrt{\left\{ (\varepsilon_z - \varepsilon_\theta)^2 + (\varepsilon_\theta - \varepsilon_r)^2 + (\varepsilon_r - \varepsilon_2)^2 + \frac{3}{2} \gamma_{z\theta}^2 \right\}} \tag{7}$$

The parameters ε_z and $\gamma_{z\theta}$ are usually those being controlled and measured during tension–torsion tests, and the values of ε_θ and ε_r can be evaluated as

$$\varepsilon_\theta = \varepsilon_r = -v^*\varepsilon_z \tag{8}$$

Here v^* is Poissons ratio corresponding to the achieved level of strains. Substituting equation (8) into equation (7) gives

$$\gamma_{oct} = 0.816 \sqrt{\left\{ \frac{4}{3} (1 + v^*)^2 \varepsilon_z^2 + \gamma_{z\theta}^2 \right\}} \tag{9}$$

The data from Fig. 1(b) for a given fatigue life N_f^* can be represented on a polar plane defined by variables r and β ($\beta = \text{arctg}(\xi)$) (Fig. 2). Any given strain state, ξ, is uniquely defined by a radial line on this plane. Thus all points which represent torsion tests lie on the line with $\beta = 135$ degrees. All strain states for push–pull are described by a curve $\beta = \beta^* = 180 - \text{arctg}(v^*)$. The

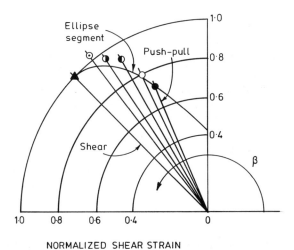

Fig 2 A polar diagram of the variation of the normalized shear strain with the strain state ξ for a given fatigue life N_f^*

points shown on Fig. 2 describe the changes of the parameter r with the strain state for the given life N_f^* shown in Fig. 1(b). A similar result can be obtained for any number of cycles to failure. It is considered here that, for a given fatigue life N_f^*, the experimental points correlate with a segment of an ellipse having one axis coincident with the line $\beta = 135$ degrees (the shear line). Then changes of the parameter with strain state can be described by a function

$$r(\xi) = b/\sqrt{(b^2 \cos^2 \bar{\beta} + \sin^2 \bar{\beta})} \qquad (10)$$

where $\bar{\beta} = \beta + \pi/4$ and b is the second (minor) axis of the ellipse which could be determined from the condition that, for uniaxial loading when $\beta^* = 180 - \text{arctg}(v^*)$, the value of the parameter r is equal to the value for push–pull, r^{pp}. This condition gives

$$b = r^{pp} \sin \bar{\beta}^*/\sqrt{\{1 - (r^{pp} \cos \bar{\beta}^*)^2\}} \quad \text{where} \quad \bar{\beta}^* = \beta^* + \frac{\pi}{4} \qquad (11)$$

It is seen from Fig. 1(b) that the parameter r^{pp} depends on the fatigue life, so the values of b and $r(\xi)$ are also dependent on lifetime.

The function $\Psi(\xi, N_f)$ in terms of equation (3) can then be obtained as

$$\Psi(\xi, N_f) = \sqrt{\left\{ \frac{b^2(N_f) \cos^2 \bar{\beta} + \sin^2 \bar{\beta}}{b(N_f)} \right\}} \qquad (12)$$

It is obvious that the value of the above function for the case of shear ($\xi = -1$) is always equal to one. For in-phase loading when λ is a constant value during a loading cycle we can assume that ξ is also a constant. Thus for a given fatigue life the value of the function is constant and equation (5) reduces to

$$\bar{\gamma}_{\text{oct}}(\xi, N_f) = 4\Psi(\xi, N_f)\gamma_{\text{oct}}^* \qquad (13)$$

where γ_{oct}^* is the octahedral shear strain amplitude.

This approach has been applied to the results quoted in references (5)–(9). The function $r^{pp}(N_f)$ in all cases has been assumed to be linear. The values of Poissons ratio v^* used in calculations has been determined in accordance with the procedure suggested in (7)(13) where the cyclic stress strain curve was available. Otherwise it has been assumed equal to its plastic value of 0.5.

The results of these calculations for the modified octahedral shear strain are shown in Figs 3–7, each for a different material. It is seen that the proposed parameter gives a good correlation for both in-phase and out-of-phase loading conditions.

Conclusions

(1) A simple criterion has been proposed for fatigue life prediction of materials subjected to combined tension–torsion. The criterion gives good correlation with the experimental values for different materials both for in-phase

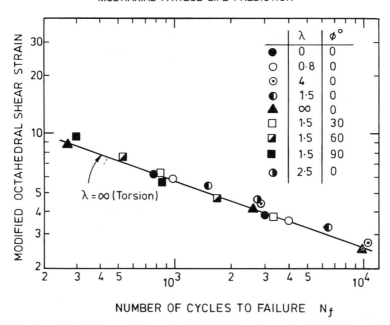

Fig 3 Tension–torsion fatigue endurance data for En15R steel tested under in-phase and out-of-phase loading conditions as a function of the modified octahedral shear strain – data from (7)

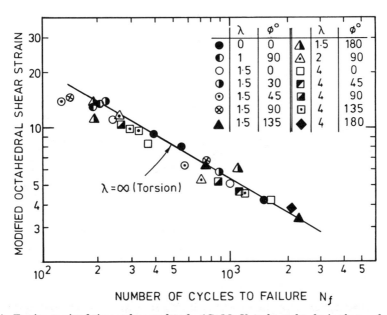

Fig 4 Tension–torsion fatigue endurance data for 1Cr-Mo-V steel tested under in-phase and out-of-phase loading conditions as a function of the modified octahedral shear strain – data from (8)

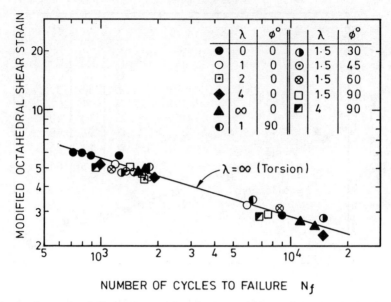

Fig 5   Tension–torsion fatigue endurance data for 316 stainless steel tested under in-phase and
out-of-phase loading conditions at a temperature of 823 K as a function of the modified
octahedral shear strain – data from (5)

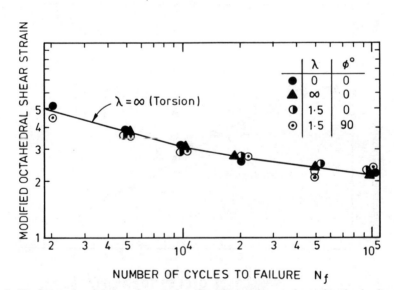

Fig 6   Tension–torsion fatigue endurance data for 321 stainless steel tested under in-phase and
out-of-phase loading conditions as a function of the modified octahedral shear strain – data
from (9)

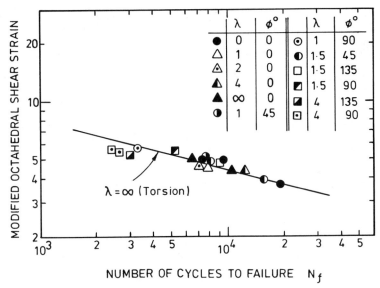

Fig 7 Tension–torsion fatigue endurance data for AISI 316 stainless steel tested at ambient temperature under in-phase and out-of-phase loading conditions as a function of the modified octahedral shear strain – data from (6)

and out-of-phase loading. The only information required for the prediction is the separate endurance curves for reversed tension–compression and for torsion.

(2) The proposed approach has taken into account the non-proportionality of the deformation process, and the dependence of the equivalent strain parameter on lifetime which may be related to different physical processes of fatigue damage at the appropriate lifetime.

Acknowledgements

Dr A. V. Troshchenko (Associate Professor from the Kiev Polytechnic Institute, Kiev, Ukraine) wishes to thank SIRIUS for arranging his visit to the University of Sheffield and the British Council for the award of a Scholarship which allowed this investigation to be carried out.

References

(1) *Biaxial and multiaxial fatigue*, EFG3 (Edited by M. W. Brown and K. J. Miller), 1989, Mechanical Engineering Publications, London.

(2) *Fatigue under biaxial and multiaxial loading* ESIS 10, (Edited by K. Kussmaul, D. McDiarmid, and D. Socie) 1991, Mechanical Engineering Publications, London.

(3) *Multiaxial fatigue*, *ASTM STP 853* (Edited by K. J. Miller and M. W. Brown) 1985, Philadelphia.

(4) MILLER, K. J. (1991) Metal fatigue – past, current and future, The John Player Lecture, Proc Inst mech. Engrs, Part C, **205**, 291–304.

(5) ANDREWS, R. M. and BROWN, M. W. (1989) Elevated temperature out-of-phase fatigue behaviour of a stainless steel, *Biaxial and multiaxial fatigue*, EGF3 (Edited by M. W. Brown and K. J. Miller), Mechanical Engineering Publications, London, pp. 641–658.

(6) BROWN, M. W. Tension–torsion experimental data for AISI 316 stainless steel under in-phase and out-of-phase loading conditions (unpublished material).

(7) FERNANDO, U. S., BROWN, M. W., and MILLER, K. J. (1991) Cyclic deformation and fatigue endurance of En15R steel under multiaxial out-of-phase loading, *Fatigue under biaxial and multiaxial loading*, ESIS 10 (Edited by K. Kussmaul, D. McDiarmid, and D. Socie), Mechanical Engineering Publications, London, pp. 337–356.

(8) KANAZAWA, K., MILLER, K. J., and BROWN, M. W. (1979) Cyclic deformation of 1% Cr–Mo–V steel under out-of-phase loads, *Fatigue Engng Mater Structures*, **2**, 217–229.

(9) SONSINO, C. M. and GRUBISIC, V. (1985) Fatigue behaviour of cyclically softening and hardening steels under multiaxial elasto-plastic deformation, *Multiaxial fatigue, ASTM STP 853* (Edited by K. J. Miller and M. W. Brown), Philadelphia, pp. 586–605.

(10) BROWN, M. W. and MILLER, K. J. (1973) A theory for fatigue failure under multiaxial stress conditions, *Proc Inst mech. Engs*, **187**, 745–755.

(11) HAVARD, D. G. and TOPPER, T. H. (1971) A criterion for biaxial fatigue of mild steel at low endurance, Proceedings of the First International Conference on Structural Mechanics in Reactor Technology, pp. 413–432.

(12) JORDAN, E. H. (1982) *Fatigue-multiaxial aspects, pressure vessels and piping: design technology 1982. A decade of progress* (Edited by Zamrik and Dietrich), ASME, New York. pp. 507–518.

(13) *ASME boiler and pressure vessel code*, Case No. 47–22 (1984) ASME, New York.

V. Kliman*

Prediction of Random Load Fatigue Life Distribution

REFERENCE Kliman, V., **Prediction of the random load fatigue life distribution,** *Fatigue Design*, ESIS 16 (Edited by J. Solin, G. Marquis, A. Siljander, and S. Sipilä) 1993, Mechanical Engineering Publications, London, pp. 241–255.

ABSTRACT The paper presents a method which enables a probabilistic approach to fatigue-life estimation under random loading. Based on a suitable segmentation of a sufficiently long random load process record and a probabilistic interpretation of material characteristics expressed by their confidence intervals, a set of life data is obtained which enables computation of a fatigue life distribution function (DF). The calculation method uses the fatigue damage accumulation hypothesis which considers the cyclic material properties, the Rainflow method, and a new method for studying the mean stress effect. This DF respects the random character of the interacting material and loading parameters, and makes it possible to determine the result-ant life for the required probability of failure occurrence – premature fracture. The method also permits the fatigue life under non-stationary random loading to be assessed. The DF obtained is in a good agreement with the experimental results.

Introduction

The majority of the service loading processes is random in character and the question of computational evaluation of the fatigue life of structures with such a loading character is still topical. This is because: (a) every part cannot be tested in a laboratory or directly in service due to time and economical or technical reasons; and (b) there are still a number of problems and require-ments in this field that must be solved. The generally accepted procedure of life prediction at a random loading consists of transforming the random process $\sigma(t)$ or $\varepsilon(t)$ (time record of stress or strain) into a macroblock of harmonic cycles (Fig. 2). This then represents the random loading process and, together with the material characteristics, it also represents the input into the hypothe-sis of fatigue damage accumulation (FDAH) on which the life is then calcu-lated.

The main problems affecting the results at this procedure can be sum-marized as follows:

(1) selection of the transformation method of the random loading process into a macroblock;

(2) length of the loading process record, from which the macroblock is obtained;

* Institute of Materials and Machine Mechanics, Slovak Academy of Sciences, Račianska 75, 830 08 Bratislava 38, Slovakia.

(3) FDAH selection;

(4) way of considering the mean loading value; and

(5) the effect of amplitudes below the fatigue limit.

Despite proper consideration of the above factors, the life calculated in such a way has some drawbacks, one of which is that the stochastic character of loading is neglected. The procedure described above gives only one life, but it is obvious that, due to random loading, the life must be a random variable. It is, therefore, necessary at least to know the scope of the interval in which the fatigue life can be expected for a given random process type. The effect of loading non-stationarities and how to calculate the fatigue life at the non-stationary random load processes (time-dependent process characteristics) are both questions which the above-mentioned procedure does not solve.

Another important problem is the consideration of the random character of the material characteristics entering the calculation, since the scatter of material properties may exert a greater effect upon the life scatter than the random character of the loading process. It is, therefore, necessary to investigate the effect of both random characteristics (material and loading) on the fatigue life in the form, in order to determine the interval in which the resultant life can be expected with the required occurrence probability.

This paper deals with the above-mentioned problems and presents a new, easy-to-use method which enables the distribution function of fatigue life (FLDF) to be predicted and considers the above-mentioned requirements.

Problem analysis

Due to the stochastic character of the loading process $\sigma(t)$, its realization (a new realization of the random process with the same statistical parameters represented by the standard deviation, probability density function, power spectral density, and the mean value) will lead to a different fatigue life value. This can be well illustrated on the basis of the energy criterion (Fig. 1) of the fatigue strength, according to which the fatigue fracture occurs after the material has absorbed a critical amount of the hysteresis energy, $W_f = \Delta W N_f$ (ΔW is the area of the hysteresis loop-work of plastic deformation amplitude per one cycle). The total hysteresis energy to fatigue fracture W_f is not a material constant, as originally believed, but a function of loading amplitude, σ_a (2). W_f increases with decreasing amplitude, i.e., with increasing number of cycles to fracture N_f. This dependence for load with $\sigma_a = $ const. can be approximated by a straight line (1) in logarithmic coordinates (Fig. 1). The accumulation of fatigue damage or the increment of absorbed energy under loading by different realizations of the $\sigma(t)$ can be schematically represented on this type of plot.

Figure 1 shows that the different value of the total hysteresis energy to fracture W_f is attributed to each process realization $\sigma(t)$. W_f and the fatigue life

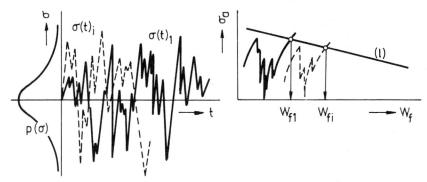

Fig 1 Hysteresis energy to fracture at a random loading

are then random variables and the scope of interval in which W_f can be expected will depend on the probability density function $p(\sigma)$ and on the slope of the boundary line. The result of the life calculation cannot, therefore, be one value but a set of values, which can be interpreted probabilistically by a distribution function.

In order to obtain the result in the required form, it would be necessary to proceed in such a way, that the fatigue life is calculated for j realizations of the random process $\sigma(t)$ by which a set of life values is obtained thus enabling the distribution function to be computed. Since this is both impractical and time consuming, the following procedure is proposed. The process record $\sigma(t)$ is divided into j number of mutually overlapping sections or segments. Each segment of the $\sigma(t)$ record then represents one random process realization. In this way the required set of lives is obtained from one long and representative record of $\sigma(t)$. The procedure is schematically shown in Fig. 2.

Materials characteristics

The fatigue–life curves at a harmonic loading are considered for the representative material characteristics

$$\sigma_a = \sigma_f'(2N_f)^b \tag{1}$$

and

$$\varepsilon_{ap} = \varepsilon_f'(2N_f)^c \tag{2}$$

Their parameters σ_f', b, ε_f', c represent the cyclic properties of material used in calculation.

Loading characteristics

A sufficiently long and representative loading process for the required service conditions is usually obtained by the strain gauge measurement as the time record of total strain $\varepsilon(t)$, or in the form of $\sigma(t)$, which is obtained either by

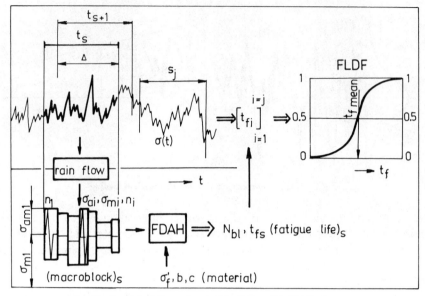

Fig 2   Calculation scheme of the fatigue life distribution

measurement of the time-dependent loading force or by the transformation of the $\varepsilon(t)$ process. The term 'represesive and sufficiently long' means that the process must include all significant service operations and that the record length must provide enough segments (process realizations $\sigma(t)$) to provide sufficient calculated sets of lives for computing the DF. Each process segment $\sigma(t)$ must be then transformed on a macroblock of harmonic cycles, which will then enter the calculation as a loading characteristic. The macroblock consists of a set of harmonic cycles with different σ_{am} amplitudes with the appropriate mean values σ_m – this represents a random process of length t_s, equal to the segment length. The Rainflow method should be used for transformation. The appropriate algorithm is mentioned in e.g. **(6)**. The result of loading processing is thus a set of j macroblocks.

Fatigue damage accumulation hypothesis and life calculation

The FDAH presented in **(4)** is used for the life calculation. According to this hypothesis the fatigue damage is characterized by plastic deformation. It has been shown here that the Palmgren–Miner's and Corten–Dolan's hypotheses are special cases of the FDAH **(4)**. The fatigue damage from one macroblock, i.e., that caused by one process segment $\sigma(t)$ will be **(4)**

$$D = \sum_{i=1}^{i=r} \frac{n_i}{N_{f\,min}} \left(\frac{\sigma_{ai}}{\sigma_{a\,max}} \right)^{(b+c)/b} \tag{3}$$

where n_i is the number of cycles with σ_{ai} amplitude, $\sigma_{a\,max}$ is the maximum amplitude in a macroblock, $N_{f\,min}$ is the number of cycles to fracture at loading with $\sigma_{a\,max}$, equations (1), and r is the number of loading levels in a macroblock. Equation (3) is derived for the macroblock of amplitudes with zero mean stress.

Fatigue life expressed in the number of macroblocks is then

$$N_{bl} D = 1 \Rightarrow N_{bl} = D^{-1} \tag{4}$$

For the life transformation into time region

$$t_f = D^{-1}t_s, \tag{5}$$

where t_s is the length of the process segment $\sigma(t)$ from which the macroblock was obtained. In this way, the life for each segment is calculated, resulting in a set of j random life values. By a statistic processing of this set, FLDF is obtained (Fig. 2) and the result is defined probabilistically.

Effect of mean stress
The σ_m value affects the fatigue life significantly, especially under stress-controlled loading. Several models which consider the effect of σ_m have been suggested. Further details and a comparison can be found in (7). The following are often considered as the most suitable procedures: (a) Morrow's approach (5) which involves reducing the fatigue strength coefficient σ'_f by the σ_m value; (b) using the parameter $(\sigma_{max} \cdot \Delta\varepsilon/2)$, designated as SWT (8); and (c) developing (a) and (b) as by Bergman (1), who modified the SWT parameter into the form $\{(\sigma_a + k\sigma_m)\Delta\varepsilon/2\}$.

In a macroblock the mean stress σ_m is attributed to each cycle. Therefore, in the life calculation according to equations (3)–(5) it is necessary to proceed in such a way that each macroblock of amplitudes with the mean values will be first transformed to a macroblock of amplitudes with zero mean values, in accordance with a method of considering the effect of σ_m. Equal fatigue life is the transformation criterion. In considering the effect of σ_m according to Morrow (5) (reduction of the coefficient σ'_f by the σ_m value), using equation (1), the transformation relation between the loading amplitude σ_a (at $\sigma_m = 0$) and the σ_{am} amplitude (at $\sigma_m \neq 0$) will be as follows

$$\sigma_{am} = \sigma_a \left(1 - \frac{\sigma_m}{\sigma'_f}\right) \tag{6}$$

For our purposes the following modification of equation (6) will be made. The condition of equal life expressed by the use of hysteresis energy, Fig. 3, is

$$W_{f(\sigma_m = 0)} = W_{f(\sigma_m \neq 0)}, \quad \text{or} \quad \Delta W N_f = \Delta W_{(\sigma_m)} N_{f(\sigma_m)} \tag{7}$$

By combining the simplified Morrow's relationship for the energy per cycle $\Delta W = 3\sigma_a \varepsilon_{ap}$ with equations (1), (2), and combining $\Delta\omega_{(\sigma_m)}$ with equation (6), after substitution into equation (7) and treatment we obtain

$$\sigma_{am} = \sigma_a \left(1 - \frac{\sigma_m}{\sigma'_f}\right)^{(c+1)/(b+c+1)} \tag{8}$$

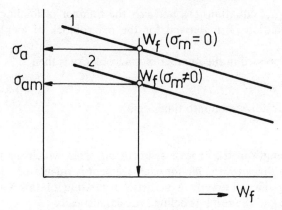

Fig 3   Total hysteresis energy to fracture as a function of stress amplitude

Equation (8) is a correction of relationship (6) by the energy criterion. According to this equation the transformation relation is also a function of material constants b, c. It should be noted that by the same procedure and with the aid of energy criterion the other models can also be modified.

Experimental verification and discussion

The method was verified experimentally on circular steel test specimens 15 mm in diameter, loaded by a random process, $\sigma(t)$, Fig. 4, in the MTS computer-controlled fatigue machine under stress-controlled mode. Material constants, parameters of equations (1) and (2), were obtained by the statistical treatment

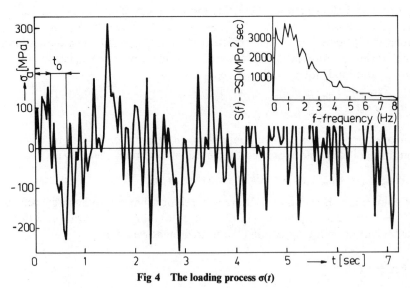

Fig 4   The loading process $\sigma(t)$

Table 1   The characteristics of test material and the results of fatigue life tests

Chemical analysis (in wt%)	C 0.242	Mn 0.44	Si 0.2	P 0.028	S 0.012	Cr 0.19	Ni 0.06
Material parameters (regr. line)	$b = -0.05295$ $\sigma_f' = 592.34$ MPa			$c = -0.52179$ $\varepsilon_f' = 0.3448$			
Exp. fatigue life (hours)	● 83.3; 93.8; 115.4 ●→ 226 unfailed						

of the experimental dependence $\sigma_a - 2N_f$ (under $\sigma_a = $ const.) and $\varepsilon_{ap} - 2N_f$ (under strain controlled tests) using the linear regression model. Fatigue tests were made on the same specimens and testing machine as in the random fatigue life tests. The experimental results of fatigue life and the characteristics of the materials used are given in Table 1. Exploiting this case let us demonstrate the proposed method and its possibilities.

Treatment of loading process

The loading process $\sigma(t)$ was recorded on a magnetic tape and after spectral analysis it was digitized at the sample frequency of 256 Hz. After digitizing a record of 614 000 ordinates ($\sim$40 min) was available. For FLDF calculation further segmentation of the $\sigma(t)$ record is necesssary.

Size of overlap value Δ and length of segment

The presented principle of FLDF calculation makes it possible to divide the record into subsequent segments, i.e., without overlapping. However, this is impractical and the overlapping should be as big as possible, since this permits a relatively large number of process segments to be obtained from a relatively short record and, moreover, the eventual non-stationarities in the loading process will be included in the life calculation. This is one of the main advantages of this procedure, since we generally do not distinguish whether a stationary or non-stationary process is concerned, and by the Rainflow method it is treated as a non-centred process. The effect of non-stationarities in the $\sigma(t)$ process will be exerted in an increased scatter of the resultant life. Size of overlapping Δ (Fig. 2) should be as big as possible and it is selected to at least 90 percent.

From the $\sigma(t)$ course in Fig. 4 it is evident that in the fatigue–life calculation according to the procedure in Fig. 2 the segment length t_s cannot be arbitrary. Very short lengths, t_0, will provide the set of lives with a great scatter and the respective DF will not be representative for the life as a consequence of a random character of $\sigma(t)$. This is reasoned by the fact that a short segment does not represent the properties of a random service process, eventually one of its realizations. Thus it is necessary to determine the minimum segment length needed for representative life calculation. It can therefore, be concluded that with the increasing segment length the scatter of calculated life will change. Hence the stabilized value of scatter of the calculated life (Fig. 5) will be the criterion for the minimum segment length.

It should be noted here that each random load process segment must have the same time length and the value of overlapping Δ must be kept constant. This guarantees that the $\sigma(t)$ process will be uniformly represented in the life calculation, i.e., none of the process part of the entire load history will be counted more times than the other.

Calculation of fatigue life distribution function

The practical procedure in life calculation is that the initial segment length is selected (in our case $t_0 = 3000$ ordinates) and at a selected overlapping Δ the distribution function and the standard deviation of life – s_{life} are calculated. The initial segment length t_0 is then increased and the life calculation is repeated until the s_{life} value stabilizes, corresponding to the minimum length $t_{s\,min}$. The result of the fatigue life calculation at segment length $t_s > t_{s\,min}$ will then be considered as representative. Figure 5 shows the results of this procedure applied to the $\sigma(t)$ process in Fig. 4, using the material constants represented by the regression line parameters (Table 1 and Fig. 7). This figure shows that the minimum necessary length $t_{s\,min} \approx 140\,000$ ordinates. For a representative calculation of FLDF a segment length $170\,000$ ordinates (~ 11 min) and a 95 percent overlap was used. From the initial record of $\sigma(t)$, 53 segments were formed from which a set of 53 macroblocks was obtained. The calculated FLDF is shown in Fig. 6, line $(KL)_r$. The effect of the mean value

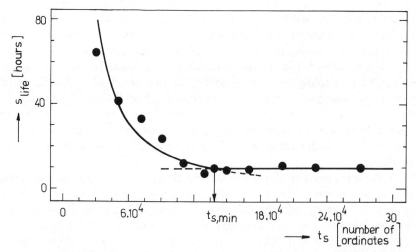

Fig 5   **Standard deviation of life in dependence on segment length of the loading process $\sigma(t)$**

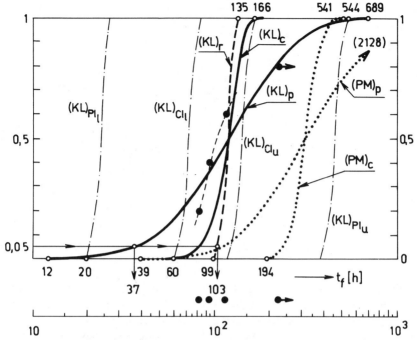

Fig 6 **Distribution function of the fatigue life at loading by $\sigma(t)$ process.**
$(KL)_r$ – **material properties represented by the regression line parameters;** $t_{f_{mean}} = 117$ h,
$s_{life} = 8.4$ h
$(KL)_c$ – **scatter of regression line parameters;** $t_{f_{mean}} = 116.5$ h, $s_{life} = 20.2$ h
$(PM)_c$ – $t_{f_{mean}} = 314$ h, $s_{life} = 52.7$ h
$(KL)_p$ – **scatter of cyclic properties of material;** $t_{f_{mean}} = 144$ h, $s_{life} = 104.5$ h
$(PM)_p$ – $t_{f_{mean}} = 392$ h, $s_{life} = 304.2$ h
$(KL)_{PI}$ – σ'_f, b, c **parameters for lower (*l*) and upper (*u*) PI limit**
$(KL)_{CI}$ – **material parameters due to lower (*l*) and upper (*u*) CI limit**
PM(KL) – **FLDF calculated based on Palmgren–Miner (Kliman (4)) FDAH**
● – **experimental life (Table 1)**

σ_m was considered according to equation (8) and for the life calculation we used the FDAH represented by equation (3). It is evident, that due to the stochaistic character of $\sigma(t)$ we can obtain the results of life for the required occurrence probability in the range 99–135 h.

The effect of material properties inhomogeneity
The inhomogeneity in the mechanical properties of the material, exerted in scatter of σ'_f, b, c, and ε'_f values, may significantly affect the course of FLDF and, therefore, it must be considered in the calculation. This will be proved for our case. For the experimental values of $\sigma_a - 2N_f$ dependence the 95 percent confidence interval (CI) for the regression line will be calculated (Fig. 7).

On evaluation of the curve parameters for the lower (CI_l) and upper (CI_u) CI limit the limit values σ'_f, b of 95 percent interval for the scatter of material

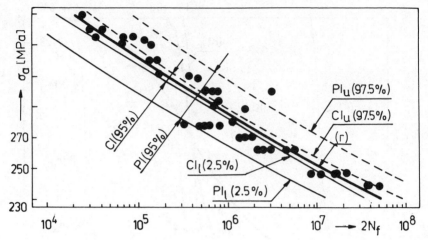

Fig 7 Experimental dependence $\sigma_a - 2N_f$. (r) – regression line; CI (95%) – 95 percent confidence interval of the regression line; PI (95%) – the confidence interval of a whole set

properties are obtained. The same must be done for the experimental data of $\varepsilon_{ap} - 2N_f$ dependence. It means that for the $(KL)_r$, Fig. 6, we can calculate the limit conditions corresponding to the material parameters for (CI_l) and (CI_u). The results are illustrated in Fig. 6. When we thus also consider the effect of material properties we can expect with 95 percent probability, the distribution function $(KL)_r$ to be in the range limited by DF for the (CI_l) and (CI_u). This depends on which σ_f', b parameters from CI for the regression line in Fig. 7 are available. In other words, the fatigue life from 60 to 166 h interval may be expected.

A similar procedure as that used for the regression line can be applied for the individual experimental points, so that in $\sigma_a(\varepsilon_{ap}) - 2N_f$ dependence we will calculate the (PI) interval, Fig. 7, where experimental result can be expected with a 95 percent probability. The DFs appropriate to the limit values of material parameters of this interval are shown in Fig. 6. The interval where, with respect to 95 percent scatter of material properties, the individual result of life under loading by $\sigma(t)$ process can be expected, falls within 20–544 h.

Fatigue life distribution function with respect to random character of loading and material properties
Based on the above, it is evident that, in our case, the scatter of material properties has a greater effect upon the life scatter than the random character of $\sigma(t)$ process. In other random processes and other materials it may have the opposite effect, therefore, such a FLDF is needed where the effect of both (material and loading) random properties is included. For this purpose, the experimental dependence $\sigma_a = f(2N_f)$ and $\varepsilon_{ap} = f(2N_f)$ for different CI and PI values is evaluated. At the selected constant value of interval change (2 percent in our case) a set of curves that represent the material properties for different

occurrence probabilities was obtained. A set of material parameters σ'_f, b, ε'_f, c representing the scatter of cyclic properties of materials corresponds to this set of curves. It is obvious here, that σ'_f and b, or ε'_f and c are the random dependent variables, because in the set of curves obtained the $b(c)$ value is attributed to each $\sigma'_f(\varepsilon'_f)$ value explicitly. The result of such a treatment of material parameters for the $\sigma_a = f(2N_f)$ dependence is shown in Fig. 8. It should be noted that the σ'_f, b or ε'_f, c values were obtained for the limits of the confidence and prediction interval, and correlation coefficient always exceeded 99 percent.

Now we have a set of material parameters and a set of macroblocks obtained from the individual process segments $\sigma(t)$. By combination of the material and loading parameters and their substitution into FDAH the set of lives is obtained from which the FLDF can be calculated, bearing in mind the random character of both the mechanical properties and that of the loading process.

Using the above-mentioned treatment, a set of 99 combinations of (σ'_f, b, c) parameters were created in such a way that $\sigma'_f = f(b)$ relationship was respected and the c parameter was attributed to (σ'_f, b) values for the same percent of the confidence or prediction interval. A set of 53 macroblocks was then available. Using equations (8), (3), and (5) a set of 5247 lives were obtained and after statistic processing, the required FLDF was acquired. The results are shown in Fig. 6.

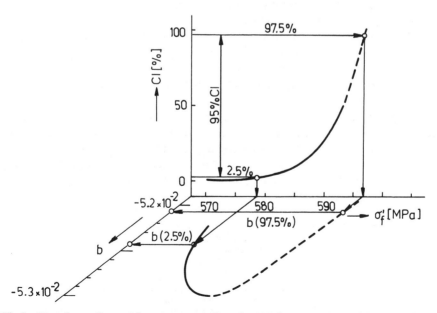

Fig 8 Dependence of material parameters σ'_f, b on the CI value; ——(---) values for the lower (upper) limit of the interval (see Fig. 7)

The procedure was applied on CI and PI, whereas the scatter of properties over the whole range of interval was considered. The results suggest that, with respect to the inhomogeneity of the material properties and the random character of the loading process $\sigma(t)$, the individual result of life can be expected to lie within the interval 12–689 h, whereas we know to attribute a life for the required occurrence probability in that interval by use of $(KL)_p$. $(KL)_c$ is FLDF due to a random character of the $\sigma(t)$ process, but only with respect to the scatter of material parameters (σ_f', b, c) representing the regression line of the experimental relationships $\sigma_a = f(2N_f)$ and $\varepsilon_{ap} = f(2N_f)$.

With respect to scatter of the regression line, the $(KL)_r$ is thus transformed into $(KL)_c$, and with respect to scatter of the whole set of material parameters the $(KL)_r$ is transformed into $(KL)_p$, (Fig. 6).

The effect of amplitudes below the fatigue limit

The number of cycles in a macroblock with the amplitudes below the fatigue limit σ_c, in comparison to the number of amplitudes above σ_c, depends on the distribution of $p(\sigma)$ function, Fig. 1, in the appropriate process segment $\sigma(t)$. The size of this ratio controls the error, which may occur in life calculation due to neglecting the amplitudes below σ_c. Following the energy criterion, only that energy which is connected with an irreversible plastic strain is considered for damaging energy. The size of the stress amplitude boundary at which the irreversible plastic strain prevails and, thus damaging energy is formed, is given by an inflection on the cyclic stress–strain curve, or $\sigma_a = f(\Delta W)$ dependence (3); this can be estimated to the value $0.8\sigma_c$. This means that on the basis of energy considerations it is advisable to consider the amplitudes below σ_c up to the limit of $0.8\sigma_c$ in the fatigue life calculation. For $\sigma_{ai} < \sigma_c$ we consider in equations (3) to (8) the curve parameters of life, equations (1) and (2), prolonged below the fatigue limit. In FLDF calculation in Fig. 6 we have considered all amplitudes below σ_c.

The differences which may be encountered in our case (Gaussian process) considering all amplitudes (life t_f) and neglecting $\sigma_{ai} < 0.8\ \sigma_c$ (life t_f^*) are shown in Table 2. $t_{f_{mean}}$, $t_{f_{min}}$, $t_{f_{max}}$ are the fatigue lives appropriate to $(KL)_p$. For this purpose the σ_c value was estimated from the equation of the cyclic stress–strain curve of this material ($\sigma_a = 761\ \varepsilon_{ap}^{0.117}$) for strain limit $\varepsilon_{apc} = (4 - 5) \times 10^{-5}$, i.e. $\sigma_c \approx 236$ MPa.

Table 2   The effect of neglecting the amplitudes below σ_c

Fatigue life (hours)	t_f	t_f^*
$t_{f_{mean}}$	144.1	145.7
$t_{f_{min}}$	12.2	12.3
$t_{f_{max}}$	689.4	696

Procedure of life estimation for ε(t) process

The loading process is very often obtained in the form of a time record of the total strain $\varepsilon(t)$ and the result of the random load process treatment is a macroblock of total strain amplitudes $\Delta\varepsilon/2$ with corresponding values of mean strain ε_m. The procedure of FLDF estimation is the same as for the $\sigma(t)$ process; however, the following FDAH should be used (4)

$$\sum \frac{n_i}{N_{f_{min}}} \left(\frac{\varepsilon_{api}}{\varepsilon_{ap_{max}}}\right)^{(b+c)/c} = 1 \tag{9}$$

In order to use equation (9) two steps are necessary.

(1) Transformation of the macroblock of total strain amplitudes $\Delta\varepsilon/2$ into a macroblock of plastic strain amplitudes ε_{ap_m} according to the equations

$$\frac{\Delta\varepsilon}{2} = \frac{(\sigma_f' - \sigma_m)}{E}(2N_f)^b + \varepsilon_f'(2N_f)^c \tag{10}$$

and

$$\varepsilon_{ap_m} = \varepsilon_f'(2N_f)^c \tag{11}$$

for the same $N_f \cdot \sigma_m$ is the mean stress corresponding to ε_m value; ε_{ap_m} is the plastic strain amplitude with the mean value ε_m.

(2) Transformation of the macroblock of ε_{ap_m} amplitudes with ε_m mean values (or corresponding σ_m values) into a macroblock of plastic strain amplitudes ε_{ap} with zero mean values. Using again the energy criterion and Morrow's approach of considering the σ_m value by the same procedure as above, the following transformation relation can be derived

$$\varepsilon_{ap} = \varepsilon_{ap_m}\left(1 - \frac{\sigma_m}{\sigma_f'}\right)^{c/(c+b+1)} \tag{12}$$

Conclusions

In a 'classical' approach to life prediction, the regression line parameters are considered as the material characteristics and the macroblock is obtained from one realization of a random process. Such a calculation yields one life value, in our case this is a random variable from the interval of 99 to 135 h (Fig. 6) depending on which process part (segment) had been used for the transformation into a macroblock. However, due to inhomogeneity of mechanical properties of a material the life may occur in a considerably wider interval, namely 12–689 h, $(KL)_p$ in Fig. 6. It is, therefore, obvious that the so-called classical approach is unsuitable and that the actual life in service, i.e., at other $\sigma(t)$ realization and a random combination of (σ_f', b, c) for PI, may differ considerably from the calculated one. Though the (12;689) hours interval is very big, these are the extreme points of interval. For a real evaluation it is to know how the life is distributed in this interval.

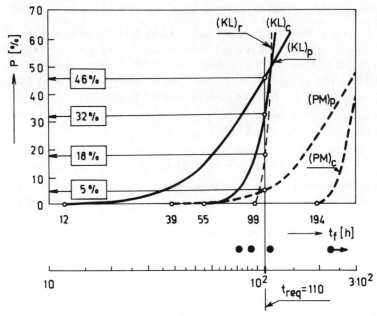

Fig 9   Fatigue failure probability with respect to required operational fatigue life according to FLDFs in Fig. 6; t_{req} – required operational fatigue life

Comparison of lives, e.g. for 5 percent failure probability (a premature fatigue failure in comparison to calculation) can show what results can be obtained when the scatter of material parameters is not considered. If we only consider the random character of the process, then from $(KL)_r$ in Fig. 6 t_f (5 percent) = 103 h. However, when the random character of material properties is considered the life t_f (5 percent) = 37 h, $(KL)_p$ in Fig. 6. This difference will naturally depend on an actual situation, i.e., on the quality of material used and the random process type, i.e., stationary or non-stationary. Moreover, it must be stressed that the result t_f (5 percent) = 37 h should be presented in such a way that it is not known what real life will be, but it will be lower than 37 h with 5 percent probability or higher than that with 95 percent probability. Thus, for life evaluation it is inevitable to employ the probabilistic approach, based on the knowledge of FLDF. Moreover, the comparison of experimental data with calculation has shown that it is insufficient just to calculate the distribution function $(KL)_c$ (Fig. 6), i.e., to consider just the scatter of regression line parameters; the distribution function $(KL)_p$ must also be used. This fact is supported by the experimental result of 226 h at the unfailed test specimen. But the main reason for using $(KL)_p$ is that the material properties of the machine part in operation are in fact represented by any experimental point in the prediction interval.

For comparison, the Palmgren–Miner FDAH has been used for calculation of the FLDF (distribution functions $(PM)_p$ and $(PM)_c$ in Fig. 6). It can be stated that the experimental results are in good agreement with the calculation and that the FLDF (KL) fits the experimental results better than FLDF (PM). The differences between individual FLDFs in Fig. 6 are more visible when the results of fatigue life calculation are related to the required value of operational life t_{req} (Fig. 9). It can be seen in our case that with respect to the required operational life the probability of fatigue failure (premature fracture) is 5 percent according to $(PM)_p$, and even zero for the distribution function $(PM)_c$. On the contrary, considering the $(KL)_c$ it is 32 percent and according to $(KL)_p$ even 46 percent. This demonstration shows how the presented procedure of life calculation can be useful for designers. The permissible value of failure probability P percent is given by requirements on safety and reliability of the structure in a given operation.

We can conclude that the definition of life again follows from the energy criterion. The boundary condition, fatigue life at loading by the cycle with σ_a amplitude, is given by the limit value of the hysteresis energy $W_f = N_f \Delta W$ (Fig. 1); ΔW is the hysteresis energy per one cycle – the area of hysteresis loop. If we consider the total life, then $W_f = W_{f, in} + W_{f, cr}$, whereas $W_{f, in}$ is the energy appropriate to the stage of life during which the ΔW is a constant and $W_{f, cr}$ is appropriate to the stage of life, where ΔW changes with an increasing number of cycles (i.e., growing crack). The significant change in loop area, from an engineering viewpoint, was observed on the test specimens at the crack which was almost 1 mm long, i.e., practically only at the beginning of the macroblock stage. The life was calculated on the basis of FDAH, equation (3), which well corresponds for the stage of life defined by use of $W_{f, in}$, i.e., with a constant area of the hysteresis loop. The calculated life can thus be characterized as the time necessary for formation of the so-called technical (engineering) initiation of the fatigue crack.

References

(1) BERGMAN, J. W. and SEEGER, T. (1979) On the influence of cyclic stress-strain curves, damage parameters and various evaluation concepts on the life prediction by the local approach, VDI – Progress Report, **18**.

(2) HALFORD, G. R. (1966) The energy required for fatigue, *J. Mater.*, **1**, 3–18.

(3) KLESNIL, M. and LUKAS, P. (1980) *Fatigue of metallic materials*, Academia, Prague, p. 240.

(4) KLIMAN, V. (1984) Fatigue life prediction for a material under programmable loading using the cyclic stress–strain properties, *Mater. Sci. Engng*, **68**, 1–10.

(5) MORROW, J. D. (1968) Fatigue properties of metals, *Fatigue design handbook*, Section 3.2, Society of Automotive Engineers, Pa, USA.

(6) NIE HONG (1991) A modified rainflow counting method, *Int. J. Fatigue*, **13**, 465–469.

(7) NIHEI, M., HEULER, P., BOLLER, Ch., and SEEGER, T. (1986) Evaluation of mean stress effect on fatigue life by use of damage parameters, *Int. J. Fatigue*, **8**, 119–126.

(8) SMITH, K. N., WATSON, P., and TOPPER, T. H. (1970) A stress–strain function for the fatigue of metals, *J. Mater.*, **5**, 767–778.

L. Gansted*

Probabilistic Chain Fatigue Model

REFERENCE Gansted, L., **Probabilistic chain fatigue model,** *Fatigue Design*, ESIS 16 (Edited by J. Solin, G. Marquis, A. Siljander, and S. Sipilä) 1993, Mechanical Engineering Publications, London, pp. 257–268.

ABSTRACT On the basis of the B-model developed in Bogdanoff and Kozin (2), a simple, numerical fatigue crack growth model incorporating the physical knowledge of fatigue is established. The model, named the Fracture Mechanical Markov Chain Fatigue Model (FMF-model), is based on the assumption that the crack propagation process can be described by a discrete space Markov theory. The model applies deterministic as well as random load. The damage, which is measured as a crack length a, is assumed to progress in steps of length δa. The damage process is described for a fracture mechanical point of view using the stress intensity factor K and Paris' equation.

Determination of the model parameter δa is based on experimental fatigue data for the given material. It can be seen from a comparison of simulated data and empirical data that both qualitative and quantitative demands are reasonably satisfied, and that the δa can be regarded as a characteristic value of the material. Since all model parameters can be assumed constant for a given material, simulations of fatigue data, and thus estimation of lifetime, can easily be performed for any structural geometry, if the geometry function is known.

Introduction

Varying loads acting on a structure will cause initiation and propagation of cracks. The cumulative damage (CD) is defined as the irreversible accumulation of damage through lifetime, which ultimately causes fatigue failure. The process is random and justifies reduction of the reliability of the structure.

Loss of human life and high economic expenses are some of the consequences caused by the risk of fatigue failure. One way to reduce the risk is to use a mathematical model which is primarily based on physically observable quantities to describe the CD process.

Usually, a distinction is made between two main groups of damage models: deterministic models and probabilistic models. Deterministic models only give information about the mean damage accumulation, thus ignoring the fluctuations characteristic of fatigue, whereas the use of a probabilistic model makes it possible to take account of the fluctuations, thus making the CD model more realistic. The fluctuations are due to variations in: (a) initial state, i.e., distribution of the initial damage, e.g., initial crack lengths; (b) magnitude and order of load cycles resulting in interaction effects in the form of retardation or acceleration of the crack growth rate; (c) and material properties in the form of e.g., inhomogeneities and loss of isotropy. All of these factors influence the fatigue crack growth.

The aim of this paper is to establish a simple, numerical fatigue crack growth model incorporating the physical knowledge of fatigue, also and to

* Department of Building Technology and Structural Engineering, University of Aalborg, Sohngaardsholmsvej 57, DK-9000 Aalborg, Denmark.

compare a series of empirical fatigue data with similar simulated data. The
simulated data are established by the model. The model, which allows results
for a given material to be transferred from one structure to another, is
described below.

The basic ideas of the FMF-model

The FMF-model introduced in Gansted *et al.* **(5)**, is briefly described below.

The FMF-model is based on the B-model, see Bogdanoff and Kozin **(2)**, i.e.,
on the assumption that the crack propagation process can be described by a
discrete space Markov theory. The discrete time is measured in numbers of
so-called duty cycles ($x = 1, 2, \ldots$, number of DCs) each consisting of a
number of load cycles, and the crack progress is described by a series of discrete damage states ($d = 0, 1, 2, \ldots, b$), where b corresponds to failure.

The damage accumulation is considered as a stochastic process in which the
possibility of damage accumulation is present every time the structure experiences a duty cycle (DC).

It is assumed that the increment of damage at the end of the DC only
depends on the DC itself and the state of damage present at the start of the
DC. Thus, the history has no influence on the increment, but is included in the
state of damage at the start of the DC. The damage only increases by one unit
at a time, see Fig. 1.

As mentioned, the damage accumulation can be regarded as a discrete-time,
discrete-state Markov process which is completely described by its transition
matrix (one for every duty cycle) and by the initial conditions.

The initial probability distribution of the damage states is given by the
vector

$$\bar{p}_0 = \{\pi_0, \pi_1, \pi_2, \ldots, \pi_{b-1}, \pi_b\}; \quad \pi_j \geq 0; \quad \sum_{j=0}^{b} \pi_j = 1 \tag{1}$$

where

$j\ = 0, 1, 2, \ldots, b-1$
$\pi_j = \text{prob } \{\text{damage is initially in state } j\}$
$\pi_b = 0$

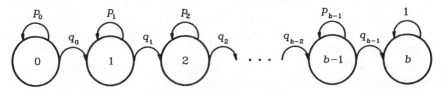

Fig 1   Illustration of the principle in the FMF model. Each circle represents a damage state, $d = 0$,
$1, 2, \ldots, b$. The p_j and q_j values are defined in equation (3)

As mentioned earlier, the damage only increases by one unit at a time. Thus, it is possible to establish the $\{(b+1) \times (b+1)\}$ transition matrix $\bar{\mathbf{P}}_i$ for the ith duty cycle given by

$$
\bar{\mathbf{P}}_i =
\begin{bmatrix}
p_0 & q_0 & 0 & 0 & \cdots & 0 & 0 & \cdots & 0 & 0 \\
0 & p_1 & q_1 & 0 & \cdots & 0 & 0 & \cdots & 0 & 0 \\
0 & 0 & p_2 & q_2 & \cdots & 0 & 0 & \cdots & 0 & 0 \\
\vdots & \vdots & \vdots & \vdots & \ddots & \vdots & \vdots & \ddots & \vdots & \vdots \\
0 & 0 & 0 & 0 & \cdots & p_j & q_j & \cdots & 0 & 0 \\
\vdots & \vdots & \vdots & \vdots & \ddots & \vdots & \vdots & \ddots & \vdots & \vdots \\
0 & 0 & 0 & 0 & \cdots & 0 & 0 & \cdots & p_{b-1} & q_{b-1} \\
0 & 0 & 0 & 0 & \cdots & 0 & 0 & \cdots & 0 & 1
\end{bmatrix}
\tag{2}
$$

$i = 1, 2, \ldots$, number of DCs
where the conditional probabilities

$$
\begin{aligned}
p_j &= \text{prob \{remain in state } j \,|\, \text{previously in state } j\} \\
q_j &= \text{prob \{go to state } j+1 \,|\, \text{previously in state } j\}
\end{aligned}
\tag{3}
$$

and

$$
\left.
\begin{aligned}
0 &\leqslant p_j \leqslant 1 \\
0 &\leqslant q_j \leqslant 1 \\
p_j &+ q_j = 1
\end{aligned}
\right\} \quad j = 0, 1, 2, \ldots, b-1
$$

$$p_b = 1$$

The damage state at the time x is then given by the vector

$$
\bar{p}_x = \bar{p}_0 \, \bar{\mathbf{P}}_1 \bar{\mathbf{P}}_2 \ldots \bar{\mathbf{P}}_x
\tag{4}
$$

where

$\bar{p}_0$ is given by (1)

$\bar{p}_x = \{p_x(j)\} = \{\text{prob \{damage is in state } j \text{ at time } x\}\}$

Once the model parameters $\pi_j, p_j, q_j; j = 0, 1, 2, \ldots, b$ are determined, the state of damage in the given structure is available at any time using equation (4). This means that all statistical information about the damage process can be represented by the model.

The problem is how to determine the model parameters. In the FMF-model this is done using a fracture mechanical point of view. This is the main point where the FMF-model differs from the B-model in which the parameters are determined solely from empirical data.

The crack length, a, is used as a damage measure which is an advantage since a is a quantity that can easily be observed. The damage is assumed to progress in steps of the length δa. Thus, the jth state of damage can be defined as

$$a_j = a_0 + j\delta_a; \quad a_j \leqslant a_b; \quad j = 0, 1, 2, \ldots, b \tag{5}$$

where

a_j = crack length at damage state j (mm)
a_0 = initial crack length (mm)
a_b = failure crack length (mm)

The damage process is described by a crack propagation equation, expressing the crack growth rate as a function of the fracture mechanical value ΔK, which is the stress intensity factor range, see **(6)**.

In agreement with the Markov assumption damage is only accumulated when the crack propagates. As long as the crack remains in a given state, the same test is repeated every time a DC is applied. This means that in a given crack state, $a = a_j$, the propagation of the crack can be modelled by a Bernoulli random variable Z, see e.g. **(1)**

$$z = \begin{cases} 0 & \text{the crack remains in the given state} \\ 1 & \text{the crack propagates } \delta a \end{cases} \tag{6}$$

The probability density function of Z is then

$$f_Z(z) = \begin{cases} p_j = 1 - q_j & \text{for } z = 0 \\ q_j & \text{for } z = 1 \end{cases} \quad j = 0, 1, 2, \ldots, b \tag{7}$$

The quantity q_j is known as the transition probability, see also equation (3). The crack growth problem is then reduced to the determination of the transition probability which is a function of the stress intensity factor range, which is a function of the crack state, i.e. $q_j = q(\Delta K_j) = q(\Delta K(a_j))$.

The most simple situation occurs if ΔK is constant, i.e., the crack tip load is constant. Thus the applied stress range $\Delta\sigma$ is decreasing, no matter how long the crack is. If so, $\Delta K_j = \Delta K$ and hence, $q_j = q$.

On the basis of the empirical Paris equation, which is one of the most frequently used crack propagation equations, and the geometric distribution, the estimation of q is possible.

Paris' equation, see **(7)**, is given as

$$\frac{da}{dN} = C(\Delta K)^m \tag{8}$$

where

da/dN = crack growth rate (mm/cycle)
C = material constant (mm/(MPa$\sqrt{m}$)^m)
m = material constant
ΔK = stress intensity factor range (MPa$\sqrt{m}$)

The crack growth rate can be estimated in different ways depending on the definition of the slope of the sample curves. Empirically, a set of sample curves – (N, a) curves – is measured. The number of duty cycles N is assumed to be observed for fixed values of the crack length a. Further, the duty cycles are assumed to be equal. In the FMF-model, the crack growth rate is then defined as the step length divided by the mean value of the number of duty cycles, $E(\delta N)$, applied in a crack state, i.e.

$$\frac{\delta a}{E(\delta N)} = \lambda C (\Delta K)^m \tag{9}$$

where

$E(\delta N) =$ the expected value of the random variable δN corresponding to the expected number of duty cycles which is used to propagate the crack one step δa

λ = number of load cycles in one duty cycle

Notice that the Paris equation has not become stochastic, all quantities in equation (9) are deterministic. (δN is a stochastic variable, but it is $E(\delta N)$ which is used in equation (9).)

Every time the crack tip is exposed to one duty cycle, the same test is repeated. The expected value and the variance of the number of duty cycles applied to propagate the crack δa is to be determined.

It is assumed that for a realization δn of δN the first $(\delta n - 1)$ duty cycles do not lead to crack propagation, i.e. $z = 0$ in equation (6), and that $z = 1$ in the δnth duty cycle, i.e., the δnth duty cycle results in crack propagation. The probability of the two events is $p^{(\delta n - 1)}$ and q, respectively. Thus, the probability distribution of δN is a geometric distribution given as

$$P(\delta N = \delta n) = f_{\delta N}(\delta n) = qp^{\delta n - 1} \tag{10}$$

The expected value and the variance of the number of duty cycles are given as the first and second moment of the geometric distribution, respectively, see **(1)**

$$E(\delta N) = \sum_{\delta n = 0}^{\infty} \delta n f_{\delta N} = \sum_{\delta n = 0}^{\infty} \delta n q (1 - q)^{\delta n - 1} = \frac{1}{q} \tag{11}$$

$$\text{Var}(\delta N) = E(\delta N^2) - \{E(\delta N)\}^2 = \frac{1 - q}{q^2} \tag{12}$$

Insertion of equation (11) into equation (9) leads to

$$q = \frac{\lambda C}{\delta a} (\Delta K)^m \tag{13}$$

This means that the transition matrix in equation (2) is determined for a given material. The damage states in structures, made of the given material, are then calculated using equation (4). The geometry of the structure is taken account of through the stress intensity factor range ΔK. Generally, ΔK is variable due to the load, but in the case of constant amplitude load (stress range $\Delta\sigma$ constant) ΔK can be assumed constant in the vicinity of the given crack position.

At every crack position, given by equation (5), the random variable δN_j, which is the number of duty cycles applied to propagate the crack one step from a_j to $a_j + \delta a$, is considered. It is assumed that a_0 is constant so that the δN_j values express the properties of the material.

As with the results for constant ΔK, the first $(\delta n_j - 1)$ duty cycles with the probability $p_j^{(\delta n_j - 1)}$ do not lead to crack propagation, whereas the δn_j th duty cycle with probability q_j results in crack propagation. Thus, the probability distribution of δN_j is a geometric distribution given as

$$P(\delta N_j = \delta n_j) = f_{\delta N_j}(\delta n_j) = q_j \, p_j^{\delta n_j - 1} \quad j = 0, 1, 2, \ldots, b - 1 \tag{14}$$

where the expected value and the variance of δN_j are given by equations (11) and (12) replacing δN and q by δN_j and q_j, i.e.

$$E(\delta N_j) = \frac{1}{q_j} \tag{15}$$

$$\mathrm{Var}\,(\delta N_j) = \frac{1 - q_j}{q_j^2} \tag{16}$$

where

$$q_j = \frac{\lambda C}{\delta a} (\Delta K(a_0 + j\delta a))^m = \frac{\lambda C}{\delta a} (\Delta K(a_j))^m \tag{17}$$

$$p_j = 1 - q_j$$

The total number of duty cycles applied to a structure to propagate the crack to the crack length a_j is

$$N_j = \sum_{k=0}^{j-1} \delta N_k \tag{18}$$

Since the random variable is a sum of independent random variables, the expected value of the number of duty cycles is

$$E(N_j) = E\left[\sum_{k=0}^{j-1} \delta N_k\right] = \sum_{k=0}^{j-1} \frac{1}{q_k} \simeq \frac{1}{\lambda C} \int_{a_0}^{a_j} (\Delta K(a))^{-m} \, da \tag{19}$$

where the sum is approximated by an integral. Notice that for $\lambda = 1$, equation (19) corresponds to the integrated Paris equation. Thus, it is possible to return to the basic point (Paris' equation) which means that the model is consistent so far.

The variance of a sum of independent variables is given as, see e.g. (1)

$$\mathrm{Var}\,(N_j) = \sum_{k=0}^{j-1} \mathrm{Var}\,(\delta N_k) = \sum_{k=0}^{j-1} \frac{1 - q_k}{q_k^2} = \frac{\delta_a}{\lambda^2 C^2} \{f(a) - g(a)\} \tag{20}$$

where the sum is approximated by an integral and where

$$f(a) = \int_{a_0}^{a_j} (\Delta K(a))^{-2m} \, da \tag{21}$$

$$g(a) = \frac{\lambda C}{\delta a} \int_{a_0}^{a_j} (\Delta K(a))^{-m} \, da \tag{22}$$

The only unknown quantity left is the step length δa, which can be estimated if the variance of N_j (or the variance of $\lambda^2 N_j$, i.e., of the number of load cycles) is known from experiments, see **(5)**.

In the case of random load, the stress intensity factor will also vary randomly and the well-known effects of acceleration and retardation might occur. The FMF-model itself does not take into account these interaction effects. This can be done using a crack closure model when ΔK is calculated, see e.g. **(8)** and **(3)**.

The damage increment in the next duty cycle will depend on the load history, the geometry of the structure and the extreme values of the duty cycle.

Introducing the effective stress intensity factor range

$$\Delta K_{eff} = \Delta \sigma_{eff} F \sqrt{(\pi a)} = (\sigma_{max} - \sigma_{cl}) F \sqrt{(\pi a)} \tag{23}$$

where

$\Delta \sigma_{eff}$ = effective stress range (MPa)
σ_{max} = maximum stress (MPa)
σ_{cl} = crack closure stress (MPa)
F = geometry function
a = crack length

the FMF-model is also available if the load is random. The only changes to be made are to replace ΔK with ΔK_{eff} in equations (8), (9), (13), (17), (19), (21), and (22).

The experimental data

The experimental data described in the following are used (a) to determine the model parameters included in the FMF-model for a mild steel, and (b) to serve as reference data for the evaluation of the applicability of the FMF-model.

The experimental data shown in Fig. 2 were obtained in 34 tests using STW22 DIN 1614 steel CCT specimens (Center Cracked Tension). The yield stress was determined as 266 MPa and the ultimate stress as 330 MPa. The CCT specimens were influenced by constant amplitude load with stress range $\Delta \sigma = 125$ MPa, maximum stress $\sigma_{max} = 127.5$ MPa, and minimum stress $\sigma_{min} = 2.5$ MPa.

In every test, the number of load cycles and the crack length (N, a) were recorded for fixed values of dN by use of the DIP technique (Digital Image Processing) between the initial crack length $a_0 = 3.0$ mm and the failure crack length $a_f = 17.5$ mm.

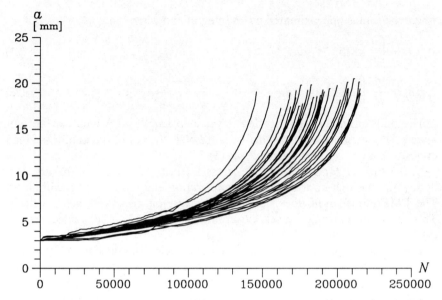

Fig 2   The thirty-four (N, a) curves obtained from the experimental data established at the Laboratory of Structural Engineering at the University of Aalborg. N = number of load cycles;
a = crack length. From Gansted (4)

On the basis of Fig. 2, the material parameters C and m in equation (17) are
found to assume the values 2.51×10^{-9} mm/(MPa$\sqrt{\text{m}})^{3.49}$ and 3.49, respectively, see also reference (4).

Simulated fatigue crack growth data

For the purpose of evaluating the applicability of the FMF-model (N, a) data
are simulated on the basis of this model. The simulations are performed by a
C programme which is developed in connection with the FMF-model. The
simulations are described below.

The main part of the model parameters used in the model was found above.
The crack step length is chosen so that the statistical properties of the simulated data correspond to the similar properties of the experimental data, see
Gansted (4). Combining equations (20)–(22) and introducing $\Delta K = \Delta\sigma\sqrt{(\pi a)}$
for the CCT specimen and assuming that $a_f \gg a_0$, the crack step length can be
estimated as

$$\delta a = \frac{C^2(\Delta\sigma\sqrt{(\pi a_0)})^{2m}(m-1)}{a_0(10^3)^m} S_{Nf}^2 \tag{24}$$

where S_{Nf} is the standard deviation of the number of cycles to cause failure for
the experimental data. Insertion of a_0, $\Delta\sigma$, C, and m from Table 1 and $S_{Nf} =$
17 115 (see Fig. 4) gives $\delta a = 0.0552$ mm.

Choosing the number of load cycles in a DC as 1, all the model parameters
are determined, see Table 1.

Table 1 **Input parameters to the simulation programme.** a_0 = **initial crack length,** a_f = **failure crack length,** $\Delta\sigma$ = **stress range,** C = **material constant,** m = **material constant,** λ = **number of load cycles in every duty cycle and** δa = **crack step length**

a_0 (mm)	a_f (mm)	$\Delta\sigma$ (MPa)	C $(mm/(MPa\sqrt{m})^{3.49})$	m	λ	δa (mm)
3.0	17.5	125	2.51×10^{-9}	3.49	1	0.0552

The number of data set in a simulation series is 500, but to keep the overview only 100 curves are shown in Fig. 3.

Comparison of the experimental data and the simulated data

The applicability of the FMF-model is evaluated below, on the basis of two criteria concerning the properties of the model. Firstly, the model must be able to give a good qualitative description of experimental fatigue crack growth data, i.e., the model should replicate the experimental curves. Secondly, the crack growth data obtained from the model must have the same statistical properties as the experimental data in the form of mean values, standard deviations, and probability density functions.

The qualitative demand is tested by plotting the fatigue crack growth curves obtained from the experimental data and from the simulated data, respectively. The form and the progress of the curves must be similar.

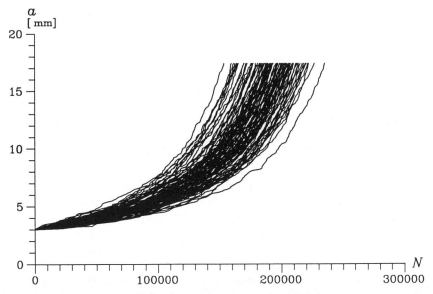

Fig 3 **(N, a) curves obtained by simulations on the basis of the FMF model with model parameters as given in Table 3.** N = **number of load cycles:** a = **crack length**

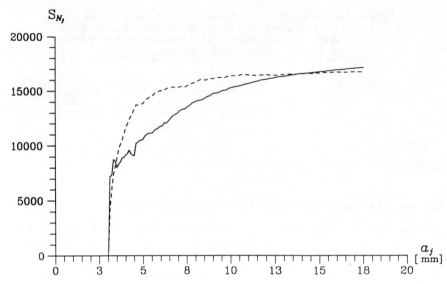

Fig 4 **Standard deviation of the number of load cycles (S_{Nj}) applied to reach the crack state (a_j), for the experimental data and the simulated data, respectively.** $j = 0, 1, 2, \ldots, 145$; $a_0 = 3$ mm; and $a_{145} = 17.5$ mm

The fatigue crack growth curves are shown in Figs 2 and 3. It is seen that the simulated data have a less smooth progress than the experimental data, but almost the same form. Thus, a small initial crack growth rate results in a large number of cycles to cause failure and vice versa. Further, the curves approach a vertical line for large a values corresponding to only a few cycles in the last part of the fatigue failure. The influence of the fluctuations of the simulated curves on the statistical values and thus on the applicability of the FMF-model, will be investigated in the near future.

The statistical values of the number of cycles applied in every crack state to reach the successive state and of the total number of cycles to reach a given crack state, i.e., the values of equations (15), (16), (19), and (20), are calculated by a statistical programme developed in connection with the FMF-model.

The total number of cycles to reach the failure state N_f and the standard deviation of the total number of cycles to reach a crack state $S_{Nj} = (\text{Var } [N_j])^{1/2}$ are considered to be the most important statistical properties, so only these will be discussed. In observing a given crack length, the primary requirement is to be able to predict the number of load cycles which can be further applied before the failure state (or any other state) is reached.

The main part of the experimental curves is concentrated within the interval $N_f \in (140\,000; 215\,000)$ cycles, whereas, for the simulated data, $N_f \in (150\,000; 220\,000)$ cycles, i.e., a small displacement but a large overlap, see Figs 2 and 3. The mean values of N_f are $185\,353$ cycles and $196\,704$ cycles, respectively, i.e., the FMF-model overestimates the lifetime by approximately 6 percent.

The progress of S_{Nj} is shown in Fig. 4 for the experimental data and the simulated data, respectively.

The standard deviation of the number of load cycles S_{Nj} applied at a crack state a_j increases for increasing crack length, see Fig. 4. The rate of increase is largest for small crack lengths ($a < 7$ mm) after which the standard deviation slowly increases towards a constant value of $\approx 17\,000$. This is most clearly seen for the simulated data. Furthermore, the absolute value of the standard deviation for the simulated data for small crack lengths is larger than for the experimental data, which can also be seen from Figs 2 and 3.

The remaining statistical curves are shown in Gansted (4).

Conclusions

The FMF-model, is a numerical cumulative damage model based on fracture mechanics in which the cumulative damage is described by a discrete-time, discrete-state Markov process. The time is measured in number of duty cycles, whereas the state of damage is given as a crack length. The crack is assumed to propagate in steps of the length δa.

On the basis of the model parameters determined for a mild steel, simulation of fatigue crack growth data has been performed. It is assumed that the crack step length δa can be regarded as a characteristic value of the material, which explicitly can be calculated from the statistical properties of the fatigue crack growth data.

The evaluation of the FMF-model is based on a comparison of the experimental data and the simulated data. Both qualitative and quantitative criteria have been used.

The qualitative demands concerning similarity in the form and progress of the fatigue crack growth curves are satisfactorily fulfilled. Thus, the simulated curves reflect the experimental curves including the interval for number of cycles to cause failure (N_f).

The comparison of the statistical analyses of the experimental data and the FMF data, respectively, shows that almost the same properties and characteristics apply. For the given choice of model parameters, the FMF-model overestimates the lifetime by approximately 6 percent and the standard deviations, S_{N_f}, only coincide for large a values. Otherwise, satisfactory results have been obtained.

A better agreement might be obtained by making it possible for the crack to progress in steps which are a multiple of the crack step length δa. This possibility will be open to further research.

References

(1) BENJAMIN, J. R. and CORNELL, C. (1970) *Probability, statistics, and decision for civil engineers*, McGraw-Hill, New York/London, p.223.
(2) BOGDANOFF, J. L. and KOZIN, F. (1985) *Probabilistic models of cumulative damage*, John Wiley, New York, Chapter 2.

(3) CORBLY, D. M. and PACKMAN, P. F. (1973) *On the influence of single and multiple peak overloads on fatigue crack propagation in 7075–T6511 aluminum, Engng Fracture Mech.*, **5**, 479–497.

(4) GANSTED, L. (1991) *Analysis and description of high-cycle stochastic fatigue in steel*, PhD Thesis, University of Aalborg, Denmark.

(5) GANSTED, L., BRINCKER, R., HANSEN, L. and PILEGAARD (1991) Fracture mechanical markov chain crack growth model, *Engng Fracture Mech.*, **38**, 475–489.

(6) HELLAN, K. (1985) *Introduction to fracture Mechanics*, McGraw-Hill, New York/London.

(7) PARIS, P. C. and ERDOGAN, F. (1963) A critical analysis of crack propagation laws, *J. Bas. Engng*, **85**, 528–534.

(8) SCHIJVE, J. (1979) Four lectures on fatigue crack growth, *Engng Fracture Mech.*, **11**, 167–221.

M. Nakajima*, H. Kunieda*, and K. Tokaji†

A Simulation of Corrosion Fatigue Life Distribution in Low Alloy Steel

REFERENCE Nakajima, M., Kunieda, H., and Tokaji, K., **A simulation of corrosion fatigue life distribution in low alloy steel,** *Fatigue Design*, ESIS 16 (Edited by J. Solin, G. Marquis, A. Siljander, and S. Sipilä) 1993, Mechanical Engineering Publications, London, pp. 269–281.

ABSTRACT A Monte Carlo simulation of corrosion fatigue life distributions has been conducted on a low alloy steel, SNCM439, by assuming that the scatter of fatigue life resulted from the variation in the growth characteristics of corrosion pits and fatigue cracks. The parameters used in the simulation were obtained experimentally. The results showed that the experimental distributions were expressed satisfactorily by the simulation in which the parameters obtained from fatigue tests at the same stress were used. Therefore, it was concluded that corrosion fatigue life distribution can be predicted by a Monte Carlo simulation, taking into account the statistical properties in the growth processes of corrosion pits and fatigue cracks.

Introduction

It is generally well known that the corrosion fatigue process includes the initiation and growth of corrosion pits followed by the growth and coalescence of fatigue cracks **(1)(2)**. Thus, corrosion fatigue life can be defined as the sum of lives consumed in those processes. However, the prediction of corrosion fatigue life has been successfully carried out by assuming that it consists of the growth processes of corrosion pits and fatigue cracks, because the initiation life of corrosion pits and the life after the crack coalescence usually have an extremely small fraction of the total fatigue life. This is a deterministic method of predicting corrosion fatigue life, but corrosion fatigue life distribution should be predicted by taking into account the statistical properties in the growth processes of corrosion pits and fatigue cracks since the scatter of fatigue life may result from the variation in both processes. In this study, a Monte Carlo simulation is performed to predict corrosion fatigue life distributions based on the experimental growth characteristics of corrosion pits and fatigue cracks, and the results obtained are compared with the experimental distributions.

Experimental procedures

The material used in this study is a low alloy steel, SNCM439 (AISI43 40), quenched at 880°C and then tempered at 500°C. The chemical composition (wt.%) is 0.38 C, 0.30 Si, 0.76 Mn, 0.018 P, 0.014 S, 0.08 Cu, 1.72 Ni, 0.75 Cr, 0.15 Mo, and balance Fe. The mechanical properties are 1158 MPa 0.2 percent

* Toyota College of Technology, Toyota Motor Corporation, Toyota-cho 1, Toyota 471, Japan.
† Department of Mechanical Engineering, Gifu University, 1-1 Yanagido, Gifu, Japan.

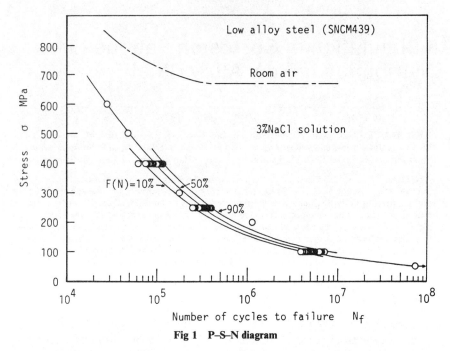

Fig 1   P–S–N diagram

offset yield strength, 1226 MPa tensile strength, 10 percent elongation, and 56 percent reduction of area.

Fatigue tests were performed in 3% NaCl solution (temperature: 30°C, pH: 5.2, dissolved oxygen: 7.82 ppm) using five cantilever type rotating bending

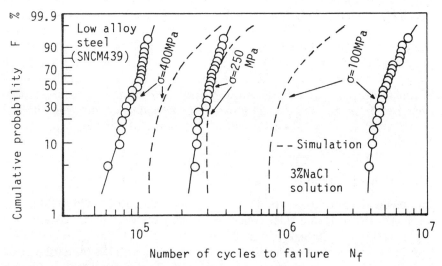

Fig 2   Experimental fatigue life distributions and predicted distributions based on the experimental results at 250 MPa (Weibull distribution)

fatigue testing machines operating at a frequency of 19 Hz. Fatigue life distributions were examined at three stress levels, and twenty specimens were allocated to each stress level. The results obtained were analysed by the three-parameter Weibull distribution.

Corrosion fatigue life distribution

The P–S–N diagram in 3% NaCl solution is shown in Fig. 1. The fatigue limit in 3% NaCl solution was approximately 100 MPa, which was considerably reduced in comparison with that in air (670 MPa). Statistical fatigue tests are conducted at 400, 250, and 100 MPa. Figure 2 indicates the corrosion fatigue life distributions at three stress levels plotted on a Weibull probability paper. As can be seen in the figure, the experimental data were well fitted by Weibull distribution functions illustrated by a solid line. The dashed line represents the simulated distributions which will be described later.

Prediction method of corrosion fatigue life distribution

Fracture surface examinations at all stress levels revealed that the growth of numerous cracks generated from corrosion pits led to final failure. This suggests that the corrosion fatigue process consists of the initiation and growth of corrosion pits followed by the growth and coalescence of fatigue cracks. Therefore, the corrosion fatigue life is defined as the sum of the lives consumed in those processes. However, the initiation life of corrosion pits and the life after the crack coalescence can be ignored, because they have an extremely small fraction of the total fatigue life. Consequently, the corrosion fatigue life, N_f, can simply be given by the following equation

$$N_f = N_{pg} + N_{cg} \tag{1}$$

where N_{pg} and N_{cg} are the growth lives of corrosion pits and fatigue cracks, respectively. Therefore, if the growth characteristics of corrosion pits and fatigue cracks are obtained experimentally, then N_f can be predicted using equation (1).

It is known that the growths of a corrosion pit and a fatigue crack are characterized as follows (3)(4)

corrosion pit growth $a_p = At^B \tag{2}$

crack growth $da/dN = C(\Delta K)^D \tag{3}$

where a_p is pit depth, t is time, da/dN is crack growth rate, ΔK is the stress intensity factor range (the maximum stress intensity factor, K_{max}, is employed as ΔK, because of the stress ratio, R, of -1), and A, B, C, and D are constants.

Regarding the transition from a corrosion pit to a crack, it is assumed that fatigue cracks are generated from corrosion pits when the stress intensity factor calculated by regarding a corrosion pit as a sharp crack exceeds a threshold value (5). Since the growth of corrosion pits and fatigue cracks is a probabilistic process having an intrinsic scatter and thus the scatter of N_f shown in Fig. 2 may result from the scatters of N_{pg} and N_{cg}, a Monte Carlo simulation of corrosion fatigue life distribution can be performed by regarding the constants in equations (2) and (3) as random variables.

Growth behaviour of corrosion pits and fatigue cracks

Corrosion pit growth

The growth behaviour of corrosion pits was monitored by replicating the specimen surface during the corrosion fatigue tests at $\sigma = 250$ MPa. Seventeen corrosion pits were selected from four specimens and were observed until the crack generated from the pits. the growth behaviour of corrosion pits is shown in Fig. 3: where corrosion pit depth, a_p, was obtained from the measurement of the surface length, $2c_p$, by assuming that the aspect ratio, $a_p/c_p = 1$. In the figure, the decreases in the growth curves are due to the scatter in the measurements by the replicating method. Based on these results, the constants A and B in equation (2) were determined by a least square method, and then plotted on a semi-logarithmic paper as shown in Fig. 4. A linear relationship between

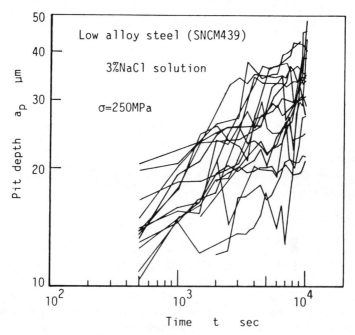

Fig 3   Growth curves of corrosion pits

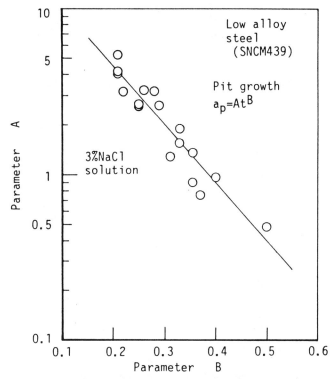

Fig 4 Relationship between constants *A* and *B*

A and *B* is obtained and can be represented by the following expression

$$A = 10^{(-3.498B + 1.353)} \tag{4}$$

In this study, it is assumed that *A* is a dependent variable and *B* is a random variable, and the probability distribution of *B* was examined. Figure 5 shows the results plotted on a log–normal probability paper, indicating that the constant *B* follows a log–normal distribution.

Transition from corrosion pit to fatigue crack

For the purpose of predicting corrosion fatigue life, the condition of the transition from a corrosion pit to a crack is required. As mentioned earlier, the transition is assumed to occur at the stress intensity factor of corrosion pits which exceeds a threshold value **(5)**. After arbitrarily selecting ten specimens of the twenty specimens allocated to each stress level, the size of corrosion pits which led to crack initiation was measured by SEM on the fracture surfaces after failure.

The size distributions of corrosion pits were plotted on a log–normal probability paper. Consequently, the surface length, $2c_{ci}$, and the depth, a_{ci},

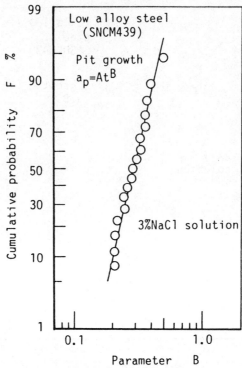

Fig 5   Distribution of constant B (log-normal distribution)

increased with decreasing stress level, for example, the pit depth ranged from 14.8 μm to 66.6 μm at 100 MPa and from 9.3 μm to 27.8 μm at 400 MPa (6). Figure 6 represents the distributions of the aspect ratio, a_{ci}/c_{ci}, plotted on a log–normal probability paper. The aspect ratio increases with increasing stress level and the average values at the stress levels of 100, 250, and 400 MPa are 0.73, 0.82, and 0.97, respectively.

Based on the above results, the stress intensity factor, K_{ci}, at which the transition occurred was calculated using the analytical solution developed by Newman and Raju (7). The results are plotted on a log–normal probability paper in Fig. 7, indicating that the K_{ci} values increase with increasing stress level. The average K_{ci} values at the stress levels of 100, 250, and 400 MPa are 0.71, 1.48, and 1.74 MPa$\sqrt{m}$, respectively. These values will be used in the simulation.

Crack growth

The crack growth behaviour was examined at $\sigma = 250$ MPa using ten specimens with a small hole of 0.1 mm diameter and 0.1 mm depth. The da/dN − K_{max} relationships obtained are shown in Fig. 8, where crack depth, a, was estimated from the measurement of surface length by assuming that the aspect

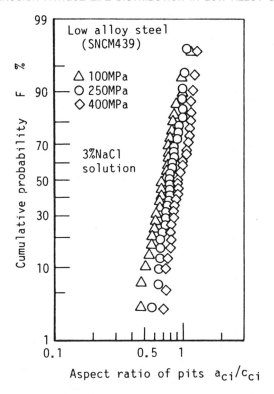

Fig 6   Distributions of aspect ratio (log–normal distribution)

ratio of cracks is unity. The constants C and D in equation (3) were determined by a least square method, and then plotted on a semi-logarithmic paper as indicated in Fig. 9. There is a good correlation between both, which is represented by the following expression

$$C = 10^{(-0.691D - 5.2996)} \qquad (5)$$

Again, it is assumed that C is a dependent variable and D is a random variable. Figure 10 shows the distribution of D plotted on a normal probability paper. As can be seen from the figure, the constant D follows a normal distribution. Finally, the final crack depth, a_f, was obtained by assuming that the failure took place at $K_{max} = 15$ MPa$\sqrt{m}$ based on fracture surface observations.

Simulation and discussion

The prediction of fatigue life distributions was conducted by a Monte Carlo simulation on the basis of the experimental growth characteristics of corrosion

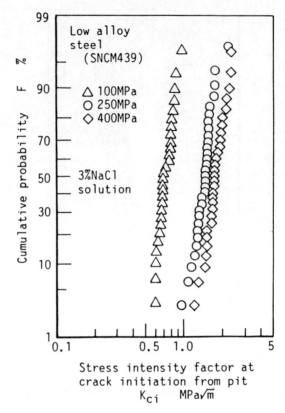

Fig 7 **Distributions of stress intensity factor at crack initiation from corrosion pit (log–normal distribution)**

pits and fatigue cracks. The procedure of the simulation is as follows.

(1) In the corrosion pit growth process, the constant B is given by the random numbers which are generated from the probability distribution of B using a computer. The constant A is calculated from equation (4).

(2) Using the average values of K_{ci} and aspect ratio, the corrosion pit depth, a_{ci}, at which cracks generate is determined.

(3) Corrosion pit growth time, t_{pg}, is calculated from equation (2), and then the corrosion pit growth life, N_{pg}, is obtained by multiplying the cyclic frequency, 19 Hz, by t_{pg}.

(4) In the fatigue crack growth process, the constant D is given by the random numbers which are generated from the probability distribution of D using a computer. Subsequently, the constant C is calculated from equation (5).

(5) Crack growth life, N_{cg} is obtained by integrating equation (3) from the initial crack depth, a_{ci}, to the final crack depth a_f.

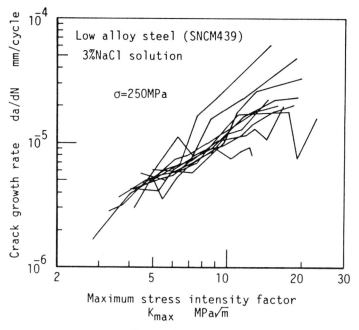

Fig 8   da/dN–K_{max} curves

(6) From steps 3 and 5, fatigue life, N_f, is determined as the sum of N_{pg} and N_{cg}.

By repeating the above procedure twenty times, the simulated data corresponding to the experimental data of twenty specimens were obtained. The results of the simulation obtained are shown by the dashed line in Fig. 2. In comparison with the experimental results, a good agreement is obtained at 250 MPa, but not at 100 and 400 MPa. In order to clarify the cause of poor agreement at these stress levels, N_{pg}, N_{cg}, and N_f at each stress level are plotted on a Weibull probability paper, and are represented in Fig. 11(a), (b), and (c). It can be seen that N_{cg} occupies a large fraction of N_f, while N_{pg} is very small. In the case of 400 MPa shown in Fig. 11(a), the predicted N_{cg} is greater than 10^5 cycles, which is longer than the experimental N_f. From the observation of the crack growth behaviour at 400 MPa, it was found that crack coalescence occurred frequently, indicating that the crack growth behaviour at this stress level is different from that at 250 MPa. Therefore, the actual crack growth life at 400 MPa may be much shorter than the prediction. Figure 11(b) shows the simulated results at 250 MPa. The cause of good agreement at this stress level is due to the appropriate use of the parameters obtained from the experiments at the same stress level. The results at 100 MPa are demonstrated in Fig. 11(c). The simulated fatigue lives are considerably shorter than

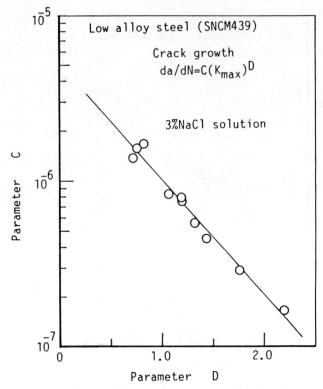

Fig 9   Relationship between constants C and D

the experimental data. From the growth experiments of corrosion pits and cracks at 100 MPa, it was confirmed that both grew much slower than those at 250 MPa.

Based on the above discussion, the difference between the experimental and the predicted distributions at 100 and 400 MPa is attributed to the stress level dependence in the growth characteristics of corrosion pits and fatigue cracks. Again, the same simulation was conducted at 100 MPa by using the parameters obtained from a small number of fatigue tests at this stress level. The results are given in Fig. 12, showing that the experimental distribution is better expressed than by the simulation using the parameters obtained at 250 MPa.

Conclusions

A Monte Carlo simulation was conducted to predict corrosion fatigue life distributions by assuming that fatigue life consisted of the growth processes of corrosion pits and fatigue cracks. The simulated results were compared with the experimental distributions. The conclusions obtained are as follows.

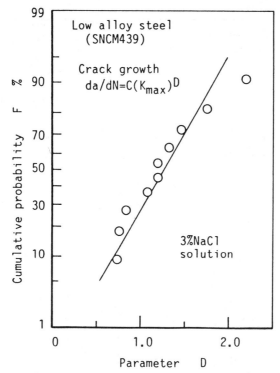

Fig 10   **Distribution of constant D (normal distribution)**

(1) A good agreement between the experimental and simulated distributions was obtained at 250 MPa, but not at 100 and 400 MPa. The cause of a poor agreement at these stresses was attributed to the stress level dependence in the growth characteristics of corrosion pits and fatigue cracks.

(2) From the simulated results, crack growth life, N_{cg}, occupied a large fraction of fatigue life, N_f, at all stress levels, while the fraction of the growth life of corrosion pits, N_{pg}, was very small.

(3) The same simulation was conducted at 100 MPa by using the parameters obtained from the fatigue tests at the same stress level. The experimental results were better expressed than by the simulation in which the parameters obtained at 250 MPa were used.

(4) Corrosion fatigue life distributions could be predicted satisfactorily by a Monte Carlo simulation taking into account the statistical properties in the growth processes of corrosion pits and fatigue cracks.

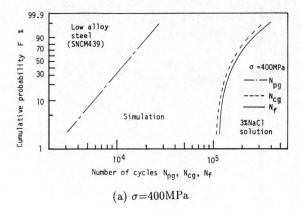

(a) $\sigma=400\mathrm{MPa}$

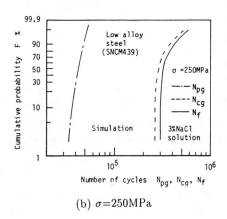

(b) $\sigma=250\mathrm{MPa}$

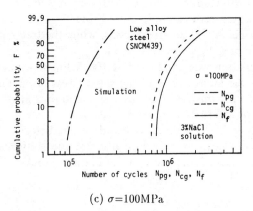

(c) $\sigma=100\mathrm{MPa}$

Fig 11   Simulated results of corrosion pit and crack growth lives (Weibull distribution)

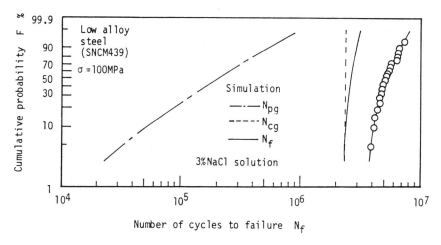

Fig 12   **Simulated corrosion fatigue life distribution based on experimental results at 100 MPa (Weibull distribution)**

References

(1) HOEPPNER, D. W. (1979) Model for prediction of fatigue lives based upon a pitting corrosion fatigue process, *ASTM STP 675*, ASTM, Philadelphia, pp. 841–870.

(2) KITAGAWA, H., FUJITA, T., and MIYAZAWA, K. (1978) Small randomly distributed cracks in corrosion fatigue, *ASTM STP 642*, ASTM, Philadelphia, pp. 98–114.

(3) ROWE, L. C. (1976) Measurement and evaluation of pitting corrosion, *ASTM STP 576*, ASTM, Philadelphia, pp. 203–216.

(4) PARIS, P. C. and ERDOGAN, F. (1963) A critical analysis of crack propagation laws, *Trans Am. Soc. Mech. Engng, Ser. D*, **85**, pp. 528–534.

(5) KOMAI, K., MINOSHIMA, K., and KIM, G. (1987) Corrosion fatigue crack initiation behaviour of 80 kgf/mm² high-tensile strength steel weldment in synthetic sea water, *J. Soc. Mater. Sci., Japan*, **36**, pp. 141–146, in Japanese.

(6) NAKAJIMA, M., KUNIEDA, H., and TOKAJI, K. (1991) Distributions of fatigue life and corrosion pit of low alloy steels, SCM435 and SNCM439, in corrosive environment, *Trans. Jap. Soc. Mech. Engrs*, **57**, pp. 2859–2865, in Japanese.

(7) NEWMAN, J. C., Jr. (1979) A review and assessment of the stress-intensity factors for surface cracks, *ASTM STP 687*, ASTM, Philadelphia, pp. 16–46.

A. Boyd-Lee* and J. King*

Discrete Statistical Model of Fatigue Crack Growth in a Ni-Base Superalloy Capable of Life Prediction

REFERENCE Boyd-Lee, A. and King, J., **Discrete statistical model of fatigue crack growth in a Ni-base superalloy capable of life prediction,** *Fatigue Design*, ESIS 16 (Edited by J. Solin, G. Marquis, A. Siljander, and S. Sipilä) 1993, Mechanical Engineering Publications, London, pp. 283–296.

ABSTRACT A discrete statistical model of fatigue crack growth in a Ni-base superalloy Waspaloy, which is quantitative from the start of the short crack regime to failure, is presented. Instantaneous crack growth rate distributions and persistence of arrest distributions are used to compute fatigue lives and worst-case scenarious without extrapolation. The basis of the model is non-material specific; it provides an improved method of analysing crack growth rate data. For Waspaloy, the model shows the importance of good bulk fatigue crack growth resistance to resist early short fatigue crack growth and the importance of maximizing crack arrest both by the presence of a proportion of small grains and by maximizing grain boundary corrugation.

Crack growth models

Previous models of fatigue crack growth can be classified into dislocation models including those based on the equation $da/dN = C(d - a)$, a stop/start model, mean long crack growth rate models, and statistical models; these are outlined below.

The fundamental dislocation models of fatigue **(1)–(5)** typically predict rapid fatigue crack growth within grains, slowing down on approaching grain boundaries and then stopping at the grain boundaries. By contrast, real cracks are observed to grow rapidly almost right up until they stop at arrests, with many arrests occurring within grains as well as at grain boundaries. These models are, therefore, currently in need of refinement to handle the more complex real situation.

Fatigue models based on the equation $da/dN = C(d - a)$ which comes from the dislocation models are further limited in that they do not make sense when they predict instantaneous crack growth rates of less than 3×10^{-10} m/cycle just before crack arrest (i.e., less than a lattice spacing per cycle) **(6)–(8)**. Possible refinements that might solve this problem are either (a) to account for the arrests more effectively or (b) to model the whole crack front rather than just a point on the crack front.

Grabowski's stop/start model **(9)** avoids these limitations of dislocation models because it is based instead on experimentally-determined crack growth

* Department of Materials Science and Metallurgy, University of Cambridge, Pembroke Street, Cambridge, UK.

rate data. The authors observed that the maximum instantaneous fatigue crack growth rate through polycrystalline Ni-base superalloys approached the maximum single crystal crack growth rate for the same material (10). Grabowski's model assumes that crack tips either grow through the material at this maximum rate or they are arrested. However, a distribution of instantaneous crack growth rates is observed experimentally and, therefore, Grabowski's model is incomplete. Whether or not this particular stop/start behaviour occurs at the lattice level has yet to be determined.

Smoothed long crack growth rate models: when long crack fronts span many grains, the scatter in the respective crack growth rate curves is mostly averaged out. Thus the mean long crack growth rate curve can be used to predict the fatigue life in this regime. This is standard commercial practice (11), but the method is unsuitable in the short crack regime where the statistical scatter is typically much greater.

The probabilistic models predict the statistical scatter of crack growth rates (12)–(18). A weakness of these models is that they lack a clear method of deriving the statistical behaviour of the fatigue process from observed crack growth rate data. A solution to this problem is presented in this paper.

Experimental

The experimental data used as input to the model described in this paper, were obtained as described in the following.

Two fatigue crack growth tests were performed at $\sigma_{max} = 880$ MPa and $R = 0.5$ on bar specimens in pure bend loading. One was in the short crack growth regime, at nominal ΔK values of 4.5–7 and 9.5–12.5 MN m$^{-3/2}$, 25 Hz sinsodual loading, the other was in the long crack regime at $\Delta K = 18.7$–19.5 MN m$^{-3/2}$, 1 Hz sinsodual loading. An example of the microstructure and one end of the long crack tip is shown in Fig. 1.

Detailed a versus N curves were obtained for each of these bands of ΔK, sufficient to give statistically significant histograms of the respective arrest persistence distributions, instantaneous crack growth rate distributions, and the probability that a growing crack becomes arrested on the next increment of fatigue crack growth. The instantaneous crack growth rate is defined here as the increment of crack growth between two successive crack length measurements divided by the number of fatigue cycles that occurred between the two measurements. The respective distributions for $\Delta K = 18.7$–19.5 MN m$^{-3/2}$ are shown in Figs 3 and 4. These two example histograms were obtained from the data marked * in Fig. 2, which was obtained from the long crack test (of one specimen). Three or more bands of detailed fatigue crack growth rate data (as in Fig. 2) are required by the model. Increments of 2000, 250, and 20 fatigue cycles were chosen for the three respective ranges of ΔK such that, on average, forty measurements were taken whilst the crack tip cut through one of the surface grains, in order that in each range of ΔK the scatter in the crack

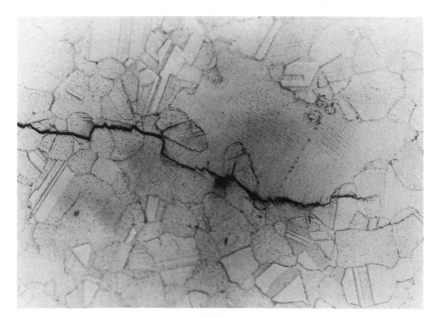

Fig 1 Micrograph of one end of the fatigue crack after long fatigue crack growth at $\Delta K = 18.7 - 19.5$ MN m$^{-3/2}$ in Waspaloy

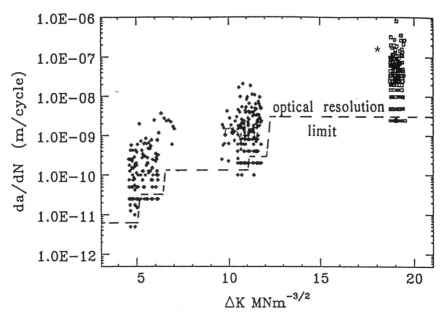

Fig 2 The fatigue crack growth rate data used as input to the model, for Waspaloy at $R = 0.5$ and $\sigma_{max} = 880$ MPa

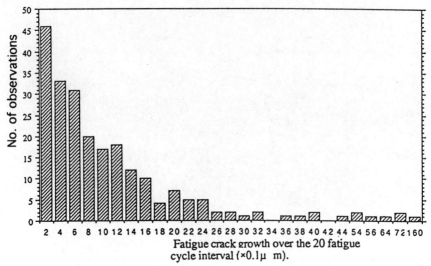

Fig 3   **Distribution of detectable instantaneous rates of long fatigue crack growth Waspaloy at $R = 0.5$ and $\Delta K = 18.7–19.5$ MN m$^{-3/2}$**

growth rates associated with microstructural events was resolved. The optical method of observation of the crack tip was used because it effectively resolved all of the scatter of the instantaneous crack growth rate curve associated with the interaction of the crack tip with the microstructure.

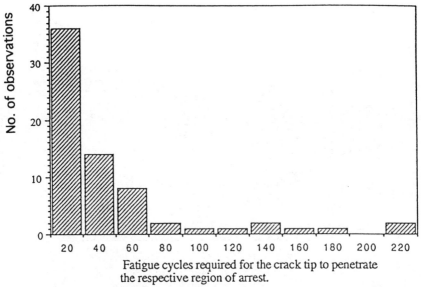

Fig 4   **Histogram of the persistence of arrests of the ends of the long crack tip in Waspaloy at $R = 0.5$ and $\Delta K = 18.7–19.5$ MN m$^{-3/2}$**

Assumptions of the model

(1) Quality control of the source material to be modelled. The source material is assumed to have a given consistent chemical composition, structure, and microstructure, manufactured into a particular component geometry, with a particular microstructural directionality if relevant, with known defect distributions, residual stress distribution(s), and manufacturing tolerances.

(2) The fatigue conditions. The experimental conditions under which the fatigue crack growth rate data used as input to the model were obtained, are the conditions for which the fatigue life predictions of the model apply.

(3) Where the fatigue cracks initiate and grow. The macroscopic and microscopic location and orientation of the initiation sites affects the fatigue crack growth behaviour.

(4) Resolution and sampling limits of observation – see the 'error in the model' section. The resolution of the crack length measurements was 0.25 microns.

(5) This model uses a measurable quantity that quantifies the progress of fatigue, for example the half surface crack length. In a different material, another measurable quantity may work with the model just as well.

(6) Reproducibility of the experimental data. The instantaneous fatigue crack growth rate distributions, the arrest distributions, and the probability that a growing fatigue crack will become arrested, are observed to be an invariant characteristic of the Ni-base superalloy being modelled at a given ΔK, subject to the above assumptions.

Eight instantaneous fatigue crack growth rate distributions have been obtained and they all have an exponential form with a mean rate which is primarily dependent on ΔK (for the given material and loading conditions). Seven out of eight of the corresponding arrest distributions also have the same exponential form, the exception being the near threshold case at $R = 0.5$ which is mechanistically different. If reproducible distributions can be obtained for a different material, then this assumption is valid for that material.

(7) Maximum crack growth rates. In the polycrystalline Ni-base superalloy U720 the authors have found that the maximum instantaneous fatigue crack growth rate is equal to the maximum instantaneous fatigue crack growth rate in single crystal U720 (10). Waspaloy is assumed to behave in a similar manner.

(8) There must be a known relationship between the crack size parameter used and the overall size of the crack.

In Astroloy, crack fronts were found to be semicircular, plus or minus a grain width; the Waspaloy in the present study is assumed to behave in a similar manner. Thus by observing the specimen surface half crack length, the overall size of the crack could be estimated.

(9) The simulation increment. The model discretizes the fatigue life into increments i of ΔN fatigue cycles. The simulation is calculated by the method of mathematical induction – i.e., the initial state of an infinite sample of cracks is known and the method of calculating the change in the distribution of crack lengths from increment to increment is known. The model assumes that four types of event occur during a fatigue cycle increment and the formulae for these changes for each set of cracks of length a–$a + \Delta a$ are all interpolated from the experimental data. They are as follows:

(a) A proportion of the cracks continue to be arrested during the current increment, according to the persistence of arrest distribution for the corresponding ΔK and the specific test conditions;

(b) The proportion of these arrested cracks that become free and then grow during the current increment is also derived from the persistence of arrest distributions.

(c) Those cracks that grow during the current increment i do so according to the instantaneous crack growth rate distribution.

(d) An experimentally determined proportion of these growing cracks will become arrested by the end of the current increment i (see Table 1).

(10) Fatigue cycle increment i, $i + 1$ independence. Slewing of the crack front causing increments of fatigue i and $i + 1$ to be dependent was observed by Lin and Cox (17)(19). However, as Fig. 5 shows, the example experimental data were not of sufficiently high resolution for this effect to be significant although

Table 1 Rate and arrest distribution information for Waspaloy

	Short $\Delta K =$ 4.5–7.5 MN $m^{-3/2}$	Short $\Delta K =$ 9.5–12 MN $m^{-3/2}$	Long $\Delta K =$ 18.7–19.5 MN $m^{-3/2}$
Arrests			
Distribution	Exponential	Exponential	Exponential
Mean persistence (fatigue cycles)	6100	570	48
Proportion of life arrested (those detected)	45%	17%	17%
(those greater than 2000 fatigue cycles	45%	0.5%	0%
Rates			
Distribution	Exponential	Exponential	Exponential
Mean rate (m/cycle)	1.56×10^{-10}	2.36×10^{-9}	3.2×10^{-8}
Mean rate excluding arrests (m/cycle)	3.03×10^{-10}	3.35×10^{-9}	5.6×10^{-8}
fatigue cycle Sampling interval	2000	250	20
P (arrest) per fatigue cycle increment	0.3716	0.2805	0.2913

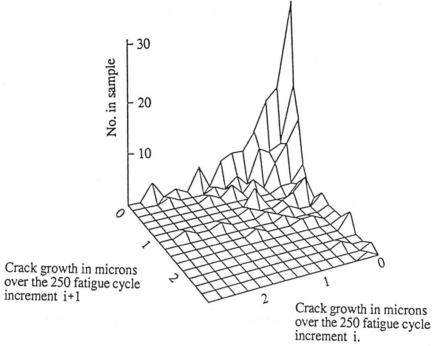

Fig 5 **Histogram of the instantaneous fatigue crack growth rate occurring during fatigue cycle increment i versus the instantaneous fatigue crack growth rate occurring during fatigue cycle increment $i + 1$ at $\Delta K = 9.5\text{–}12$ MN m$^{-3/2}$**

it was detected at $\Delta K = 18$ MN m$^{-3/2}$ after persistent arrests only. Unless the crack growth rate data are particularly detailed, this memory effect may not need to be accounted for. The current model does not take account of this effect.

(11) The ΔK limits of the model, are the same as those of the input experimental data for which statistically significant instantaneous crack growth rate and persistence of arrest distributions can be obtained.

(12) There is the generally recognised need to continue searching for any discrepancies in the behaviour of the material between the real and experimental conditions.

Description of the model

The model simulates the fatigue life, by predicting how a given infinite population of fatigue cracks, with a given crack length distribution, grow over the course of a number of fatigue cycles. The model represents the state of this infinite sample of cracks as a crack length distribution as shown in Fig. 6. This

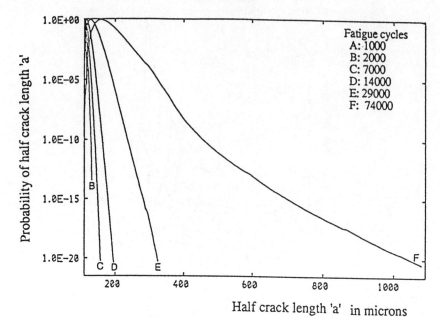

Fig 6 **Prediction of how the crack length distribution spreads out with increasing fatigue cycles from an initial infinite sample of 100 micron fatigue cracks in Waspaloy. The simulated test conditions included $\sigma_{max} = 880$ MPa and $R = 0.5$**

distribution is held in the form of a histogram in a numerical array where each element of the array holds the cracks that are of length $a-a + \Delta a$. There are several thousand elements in this array so that the respective finite element error is negligible. The initial state (or the rate at which the fatigue cracks initiate) is specified as an input to the model.

In the simplified example given in this paper, all of the fatigue cracks start at 100 μm in length. Then the simulation is advanced by an increment of ΔN fatigue cycles. The changes that occur during this increment are defined by assumption 9 in the above. In a similar manner increments 2, 3, ..., are calculated until the simulation can predict no further due to assumption (11).

Executing assumption (9) requires the following interpolation. In the example given, the persistence of arrest and instantaneous crack growth rate distributions were obtained only at three values of ΔK. However, a particular element i, containing cracks of length $a-a + \Delta a$ will be at a different ΔK according to the crack length. Thus, in general for any element i the model has to interpolate between two experimentally-determined distributions to get the respective distributions for element i. To save cpu time these distributions are calculated for each element once and for all before the simulation begins. In the version 2.0 of this model, a linear interpolation of the histograms between bands is used. The current version uses exponential interpolation which fits the data better. Also a complex band width correction (beyond the scope of

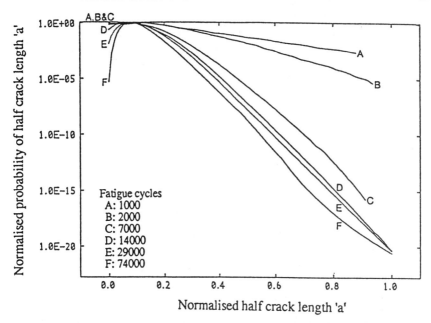

Fig 7   Normalized view of the predicted fatigue crack length distributions shown in Fig. 6. Allowing for the quantization error of the early distributions (A and B), the form of these distributions quickly equilibriates to the form C, D, E, and F

this paper) is now used to compensate for the fact that the bands of data are not measured at constant ΔK but within narrow ranges of ΔK. Also version 3.0 of the model treats the late stages of fatigue cracks growth (i.e., when there is negligible scatter in the fatigue crack growth rates) separately – these cracks grow effectively continuously along the respective a versus N curve. These recent refinements of the model are outside the scope of this paper.

An added complication is that there is an optimum fatigue cycle increment for each element, since the input experimental data has a fatigue cycle increment that varies with ΔK and hence with crack length. In practice these optimum increments have to be quantized, to substantially reduce the cpu required to solve the model. The categories of fatigue cycle increment are fixed at a factor of two apart from each other. Thus different parts of the distribution are sometimes subjected to different fatigue cycle increments as the total number of fatigue cycles steadily increases.

Once the above precalculations are complete, the fatigue simulation loop is entered. During each loop the following tasks are undertaken.

(1) Check whether the simulation is complete.
(2) Output predictions when necessary.
(3) Determine which cracks will be affected by the next increment of fatigue cycles. When some cracks have different quantized fatigue cycle increments

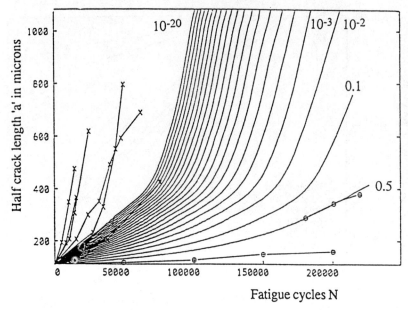

Fig 8 Another way of representing the predictions of the model as *a* versus *N* for lines of isoprobablity, compared against experimental observation.

Key:
✕——✕ Single crystal U720, σ_{max} = 890 MPa – Reed (20)
⊖——⊖ Polycrystalline Waspaloy, σ_{max} = 880 MPa
———— Predictions of model for Wasploy, σ_{max} = 880 MPa

to others, those with the smaller increment have to be accounted for more often than those with the higher increment.

(4) The simulation is restricted to the significant part of the crack length distribution.

(5) The fatigue events of the current increment are calculated from the look-up-tables.

(6) The effect of these events is added to the crack length distribution.

Results

Given the stated condition, the predicted crack length distributions at different numbers of fatigue cycles are shown in Fig. 6. The normalized distributions (Fig. 7) tend to an approximately constant form within 7000 fatigue cycles from the start of the simulation. The fine undulations of the curves predicted (Figs 6 and 7) come from the fatigue cycle increment discretization error (see description of model) but this contributes no more than a 5 percent error to the model. These undulations can be either avoided or reduced by either using less variable fatigue cycle increments or by using categories of ΔN of less than a factor of two apart from each other.

A more suitable representation of the predictions for engineers is shown in Fig. 8, which is to be interpreted as follows. The line marked 0.5 is the predicted mean a versus N line. 0.1*100 percent (i.e., 10 percent) of the cracks grow faster than the 0.1 line, 1 percent grow faster than the 0.01 line and so on. Suppose, for example, that the chosen probability of a particular worst case scenario fatigue life is 10^{-10}, that an initial defect size of 100 microns has to be assumed, and that a 1 mm crack would result in a failure, then the engineer reads off the 10^{-10} curve that the predicted life is 125 000 fatigue cycles $\pm$ the calculated error of 12 percent.

The polycrystalline crack growth data for Waspaloy used as an input to the model, is also plotted on Fig. 8 and it lies within the expected range of the predicted mean crack growth rate line. Furthermore, the predicted worst case scenario cracks approach crack growth rates that are comparable to single crystal crack growth rates, especially during the short crack regime. This is to be expected, as stated earlier, because the maximum polycrystal fatigue crack growth rates at points on the crack front approach the maximum single crystal crack growth rates. Thus the predictions of the model fit the experimental data. Single crystal crack growth rate data were not available for Waspaloy, so available data for U720 are plotted. The predictions of the model compare favourably with the form of the experimentally-determined a versus N family curves for 7475–T7351 aluminium determined by Hovey (21), although it is a somewhat different material.

Error in the model

Virtually all of the error in the model originates from the experimental data, as follows. The experimental error in the measurement of the means of the distributions can be expressed as the variance of the expected mean divided by the expected mean

$$\frac{\sigma(\bar{x})}{\bar{x}} = \frac{1}{\sqrt{n}} * 100\% \tag{1}$$

Thus the chosen sample of $n = 200$ data points per distribution adds a 7 percent variance to the model. An estimated 5 percent error is also added as a result of the quantization of the fatigue cycle increments.

Note the relatively high accuracy of predictions even though only two fatigue tests are conducted. The reason for this accuracy is that the detailed crack growth measurements are sampling the events that account for the scatter in fatigue lives. Even though there are fatigue events that are below the 0.25 micron resolution of the measurements, these are accounted for in the calculated variance of the mean of the distributions. This error analysis is based on the assumption that a growing fatigue crack tip is continually advancing through fresh microstructure.

Interpolations are used by the simulator to calculate quantities based on the input experimental data, for ΔK values other than those for which input experimental data are available. This causes shorter fatigue lives to be predicted than is actually the case. This error could either be made negligible by inputting more distributions (i.e., for more ΔK values) or the error could also be reduced by using curved rather than linear interpolations between the successive distributions.

Application of the model

The model provides a new method of making fatigue life predictions. Instead of testing a large number of specimens to failure, a few specimens are tested, taking a large number of crack length measurements during the course of each test, so that instantaneous crack growth rate and persistence of arrest distributions can be obtained for a range of ΔK values. This kind of experimental data is most easily obtained by automated crack length measurement systems **(22)**. The model then provides a powerful analysis of this data. Further work is currently being undertaken to determine how the model performs alongside existing lifing methods.

It is important that the model accounts for the statistical scatter of the crack growth rates. For example, Kendall **(23)** observed that decreasing the grain size of a Ni-base superalloy improved the short fatigue crack growth resistance. However, the model shows that in a worst case scenario, there is less crack arrest (typically associated with grain boundaries) and, therefore, the benefits of reducing the grain size are not as great as the mean crack growth rate suggests they are.

Thus the model shows that improving the bulk fatigue crack growth resistance is beneficial. For example a hypothetical 100 percent increase in the bulk fatigue resistance, i.e., reducing the maximum crack growth rates, is predicted to give a 25 percent increase in the fatigue life (from an initial crack length of 100 microns). If the assumed initial defect size is smaller than 100 microns then the bulk fatigue crack growth resistance becomes correspondingly more important. There is also the possibility that improving the bulk fatigue crack growth resistance may help to increase the persistence of the arrests.

The model predicts that some crack arrests occur in worst case scenarios, and, therefore, it is desirable to maximize the probability and persistence of arrests. This might be achieved by employing a necklace microstructure with a small minimum grain size and by developing highly corrugated grain boundaries.

The relative importance of the bulk fatigue resistance and of crack arrest in a given material can be assessed by the model. This should help the materials engineer to tune the composition and fabrication of the material for optimum performance. The difference between the single crystal fatigue life and the polycrystal fatigue life in Ni-base superalloys was shown by the model to be primarily due to arrests directly and indirectly associated with grain boundaries.

Conclusions

The conclusions of the modelling are as follows.

(1) The life predictions of the model agree well with experimental crack growth life data.

(2) The model is a useful research tool for deciding criteria for the optimization of microstructure with respect to fatigue resistance and for determining the fatigue performance of experimental materials. For example, for Ni-base superalloys good microstructural properties for maximum fatigue crack growth resistance could include:

(a) a necklace grain structure with a very small minimum grain size (in agreement with Kendall (**23**));

(b) good corrugation of grain boundaries to maximize the tortuousity of the crack path especially in the near vicinity of grain boundaries, to maximize the crack arrest contribution; and

(c) good bulk fatigue crack growth resistance (i.e., when cracks are growing – not arrested) especially during the early stages of short fatigue crack growth.

Acknowledgements

The authors would like to thank Professor C. J. Humphreys for provision of laboratory facilities in the Department of Materials Science and Metallurgy at Cambridge University. The authors are grateful to Rolls-Royce plc, to the MOD for financial support, and to Dr L. Grabowski and Dr P. A. S. Reed for useful discussions.

References

(**1**) NAVARRO, A. and DE LOS RIOS, E. R. (1987) *Fatigue Fract. Engng. Mater. Structures*, **10**, 169–386.

(**2**) NAVARRO, A. (1988) *Phil. Mag. A*, **57**, 15–36.

(**3**) NAVARRO, A. (1988) *Phil. Mag. A*, **57**, 37–42.

(**4**) NAVARRO, A. and DE. LOS RIOS, E. R. (1988) *Fatigue Fract. Eng. Mater.*, **11**, 383–396.

(**5**) NAVARRO, A. & DE. LOS. RIOS, E. R. (1990) *Phil. Mag. A*, **61** 435–449.

(**6**) HOBSON, P. D. (1982) *Fatigue Engng. Mat. Structures*, **5**, 323–327.

(**7**) YATES, J. R., ZHANG, W., and MILLER, K. J. (1993) The initiation and propagation behaviour of short fatigue cracks in Waspaloy subject to bending, *Fatigue Fract. Engng. Mater. Structures*, **16**, 351–362.

(**8**) SUN, Z., DE LOS RIOS, E. R. and MILLER, K. J. (1991) *Fatigue Fract. Engng. Mater. Structures*, **14**, 277–291.

(**9**) GRABOWSKI, L. and KING, J. E., (1992) Modelling short fatigue crack growth behaviour in Nickel-base superalloys, *Fatigue Fract. Engng. Mater. Structures*, **15**, 595–606.

(**10**) BOYD-LEE, A. D. (1990) 'Fatigue crack growth resistance and choice of short crack path in 3 Ni-base superalloys', (unpublished document).

(**11**) MILLER, K. J., SLADE, J., and WU, X. J. (1992) 'Short Fatigue Cracks', Ed. Miller, K. J., MCEP, (to be published).

(**12**) YANG, J. N., SALIVAR, G. C., and ANNIS, Jr., C. G., (1983) *Engng. Fracture Mech.*, **18**, 257–270.

(13) NEWBY, M. J. (1987) *Engng. Fracture Mech.*, **27**, 477–482.
(14) LAUSCHMANN, H. (1987) Stochastic model of fatigue crack growth in heterogenous material, *Engng. Fracture Mech.*, **26**, 707–728.
(15) IHARA, C., and MISAWA, T. (1988) *Engng. Fracture Mech.*, **31**, 95–104.
(16) COX, B. N., PARDEE, W. J., and MORRIS, W. L. (1986) 'A statistical model of intermittent short fatigue crack growth', *Fatigue Fract. Eng. Mater. Structures*, **9**, 435–455.
(17) COX, B. N. and MORRIS, W. L. (1987) 'Model-based statistical analysis of short fatigue crack growth in Ti 6Al-2Sn-4Zr-6-Mo', *Fatigue Fract. Eng. Mater. Structures*, **10**, 429–446.
(18) TANAKA, K. (1986) 'Small Fatigue Cracks', [Proc. Conf.] Santa Barbara, California, USA, The Metallurgical Society/AIME, 420 Commonwealth Dr., Warrendale Pennsylvania 15086, USA.
(19) LIN, Y. K. and YANG, J. N. (1983) *Eng. Fracture Mech.*, **18**, 243–256.
(20) REED, P. A. S. & KING, J. E. (1992) Comparison of long and short crack growth in poly-crystalline and single crystal forms of Udimet 720, *Short fatigue cracks*, ESIS 13 (Edited by K. J. Miller and E. R. de los Rios), Mechanical Engineering Publications, pp. 153–168.
(21) HOVEY, P. W., *et al.* (1983) *Engng. Fracture Mech.*, **18**, 285–294.
(22) YI, L., SMITH, R. A., and GRABOWSKI, L. 'An automated image processing system for the measurement of short fatigue crack growth', *Fatigue crack measurement techniques and applications* (Edited by Marsh, K. J. and Smith, R. A.), EMAS.
(23) KENDALL, J. M., GRABOWSKI, L., and KING, J. E. (1989) '*Materials development in turbo-machinery design*', (Edited by Taplin, D. M. R. and Knott, J. E.), Institute of Metals, pp. 231–240.

M. Kawaguchi*, Y. Yaginuma*, and V. R. Macam, Jr.†

Fatigue Experiments of Microconcrete Beams under a Running Load

REFERENCE Kawaguchi, M., Yaginuma, Y., and Macam, Jr., V. R., **Fatigue experiments of microconcrete beams under a running load,** *Fatigue Design,* ESIS 16 (Edited by J. Solin, G. Marquis, A. Siljander, and S. Sipilä) 1993, Mechanical Engineering Publications, London, pp. 297–306.

ABSTRACT Singly reinforced microconcrete beams without shear reinforcement were tested under a static load, a fixed-point repeated load, and a running load. The beams had constant cross-sections and had four variations of span lengths from 40 cm to 100 cm. The microconcrete had a maximum aggregate size of 2.5 mm and the steel ratio was 1.8 percent.

A significant reduction in the fatigue life of the beams under a running load was observed. For a given magnitude of load, the fatigue life was reduced to about $1/10^4$ of that under a fixed-point repeated load. A running load subjects the beam to a larger effective stress range resulting in more serious damage than that of a fixed-point repeated load. This reduction in fatigue life should be considered in the fatigue design of reinforced concrete bridge beams.

Introduction

Fatigue strength of reinforced concrete (RC) structures subjected to moving load attracted the attention of researchers after extensive damage of RC bridge decks was observed in Japan. Static tests and fixed-point repeated load tests performed on prototype slabs showed crack patterns different from those occurring in actual use.

Experiments on prototype slabs by Matsui (3) and Sonoda (5), and scale model slabs by Kawaguchi (1) subjected to moving loads showed crack patterns similar to the actual damage of RC bridge decks. More importantly the experiments revealed a remarkable decrease in the fatigue strength of slabs subjected to moving load as compared with those subjected to fixed-point repeated load.

Wheel loads on bridge decks are transferred to the longitudinal beams. The effect of this moving load was not given enough consideration in the fatigue design of RC beams. Many studies on the fatigue strength of beams subject to fixed point loading are reported in the literature, but hardly any research has been carried out on the relationship between the fatigue of beams under a running load and those under a fixed-point repeated load (2).

Experimental investigation

Test specimen

The beam dimensions were 5 cm wide and 8 cm high with an effective depth of 7 cm. The steel ratio was 1.8 percent. The span of the beams were 40, 60, 80,

* Nihon University, 7-24-1 Narashinodai, Funabashi City 274, Chiba, Japan.
† Technological University of the Philippines, Manila, PO Box 3171, Philippines.

and 100 cm. They were cast sideways so as to have a smooth top surface for the moving load experiment. The beam specimens were allowed to stay in the forms for 24 h. They were then removed from the forms, cured in water for ten days, and then cured in air for more than twenty-eight days before testing.

The microconcrete used was a mixture of naturally graded river sand with a maximum size of 2.5 mm and early strength portland cement. The main reinforcement used was 6 mm deformed bars with a yield stress of 393 MPa. Stirrups made of 2 mm annealed wire were placed at the furthest ends of the reinforcement to keep the bars in place. Figure 1 shows the beam dimensions and the reinforcement details.

Test methods

The beams were coded according to their span lengths, Series A, B, C, and D correspond to 40, 60, 80, and 100 cm. The prefixes S, R, and M attached to the letters A–D mean static load, repeated load, and moving load, respectively.

All beams were simply supported. For each series, some beams were tested statically to failure, while others were subjected to a fixed-point repeated load and the rest were tested under a running load.

Fixed-point repeated load was applied at a rate of 4 Hz. The load was applied through a caster wheel similar to the one used for the moving load.

For the moving load test, a specially designed machine was used (Fig. 2). The beam was set on a movable bed, which was driven back and forth by an air piston 15 times/min. The load was applied to the beam through a caster wheel

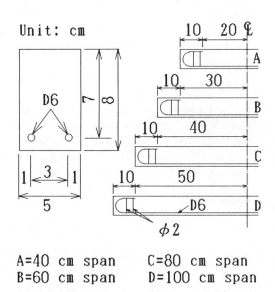

Fig 1   Beam dimensions and bar arrangement

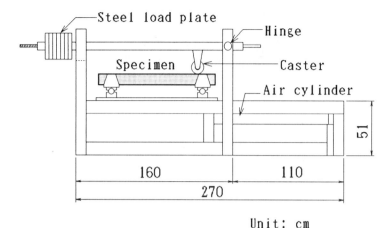

Unit: cm

Fig 2   Running load test machine

by a weight and lever. The caster, which is commercially available, had a steel wheel with a solid rubber tyre. The support condition is hinged–hinged in order to fix the beam on the bed.

Results

Static load

The results of the static load tests of materials and beams are given in Tables 1 and 2 and crack patterns of beams at failure are shown in Fig. 3. The shear strengths of the beams were calculated using equation (1) which was proposed

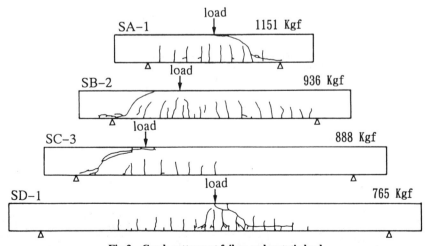

Fig 3   Crack patterns at failure under static load

Table 1 Material strength of beams

Span length (cm)	Compressive strength (MPa)	Tensile strength (MPa)
40	36.1	4.62
60	38.1	4.32
80	47.2	4.09
100	44.4	4.13

by Okamura and Higai (4). The average of the ratios of calculated to tested values is 1.11 while the coefficient of variation is 2.7 percent.

$$V = 0.20\, f_c^{1/3}(0.75 + 1.4\, d/a)(1 + \beta_\rho + \beta_d)b_w\, d \tag{1}$$

where $\beta_\rho = \rho_w^{1/2} - 1$, $\beta_d = d^{-1/4} - 1$, f_c is the compressive strength of concrete in MPa, ρ_w is the steel ratio in percent, d is the depth of the beam, b_w is the width of the web, and a is the shear span in meters.

The strength of the microconcrete beam and its mode of failure was obviously influenced by the point of application of the load. Series A, B, and C beams all failed in shear with the exception of SC-2 and SD-1 beams which failed in flexure.

For all spans, flexural cracks were initiated and diagonal cracks developed later on one side of the beam. Fractures took place along the diagonal crack which developed finally between the load and the nearest support.

Fixed-point repeated load

In this test a minimum load of 0.981 kN was applied to all the beams. The results are shown in Table 3 and the crack patterns at failure are shown in Fig. 4. The load ratio was the ratio of the maximum load to the calculated ultimate load using equation (1).

The crack patterns and failure modes of the microconcrete beams subjected to fixed-point repeated loading were practically the same as those of the beams

Table 2 Static strength of beams

Span length (cm)	Specimen	Load distance from nearest support (cm)	Ultimate load (kN)
40	SA-1	20	11.29
	SA-2	10	8.35
	SA-3	10	19.22
	SA-4	10	10.54
60	SB-1	30	10.63
	SB-2	20	9.18
	SB-3	10	12.16
80	SC-1	40	8.78
	SC-2	40	9.28
	SC-3	20	8.71
100	SD-1	50	7.5

Table 3   Fatigue strength of beams under fixed-point repeated load

Span length (cm)	Specimen	Maximum load (kN)	Loading from the support (cm)	Load ratio	Cycles to failure
40	RA-1	8.43	15	0.710	17 500
	RA-2	7.36		0.619	281 928
	RA-3	7.94		0.668	53 551
	RA-4	9.02		0.759	50 010
60	RB-1	4.90	18	0.492	252 697
	RB-2	5.88		0.591	940 578
80	RC-1	7.65	30	0.767	922
	RC-2	6.72		0.674	278 593
	RC-3	6.96		0.698	1 145

loaded statically. The beams were accompanied by longitudinal cracks along the main reinforcement in the final stage. The failure was on one side of the beam only while the other side was generally intact except for a few cracks.

Moving load

Results of the moving load tests are shown in Table 4 and Figs 5–9. The load ratio was the ratio of the running load to the calculated ultimate load of beams using equation (1). When beams failed under the running load, the distance from the top of the dominating diagonal crack to the support was typically 2–3 times the depth of the beam. Therefore, a shear span to depth ratio of 2.5 was used in the calculations.

Cracking process

The process of crack initiation and development was similar for all the beams regardless of span lengths (Figs 5–8). At the first pass of the load (one complete pass of the load on the beam is considered one cycle), flexural cracks were initiated along the length of the beam at wide spacings. As the number of

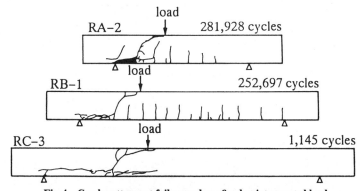

Fig 4   Crack patterns at failure under a fixed-point repeated load

Table 4   Fatigue strength of beams under a moving load

Span length (cm)	Specimen	Running load (kN)	Load ratio	Cycles to failure
40	MA-1	5.88	0.477	2 996
	MA-2	5.39	0.438	64 236
	MA-3	5.82	0.472	1 452
60	MB-1	5.88	0.591	796
	MB-2	5.39	0.541	1 658
	MB-3	4.90	0.492	104 532
	MB-4	4.90	0.492	36 506
	MB-5	4.41	0.443	138 444
80	MC-1	5.88	0.606	1 192
	MC-2	5.39	0.556	498
	MC-3	4.90	0.505	42 114
	MC-4	4.66	0.480	159 038
	MC-5	5.15	0.531	30 442
100	MD-1	4.90	0.544	1 082
	MD-2	3.92	0.435	305 978
	MD-3	4.41	0.490	61 402

passes increases, new cracks were initiated in between the old cracks, while the old cracks propagated upwards. The cracks at the outer third spans started to incline toward the midspan. Longitudinal cracks were initiated at the lower tail of the inclined crack and propagated slowly at the level of and along the

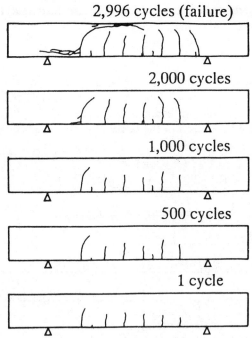

Fig 5   Crack development under a running load (MA-1 beam)

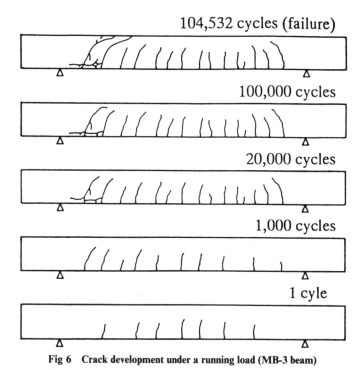

Fig 6   Crack development under a running load (MB-3 beam)

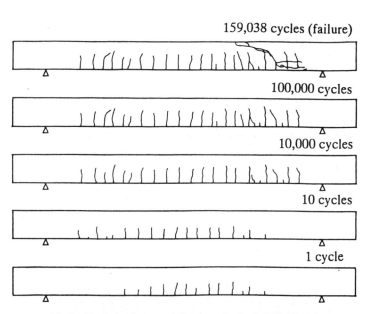

Fig 7   Crack development under a running load (MC-4 beam)

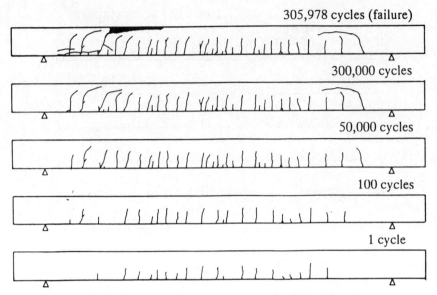

Fig 8   Crack development under a running load (MD-2 beam)

main reinforcement. As the number of passes reached the fatigue life of the beam, the inclined cracks at the outer third spans propagated upward quickly until it reached the uppermost surface of the beam. When this occurred, the longitudinal cracks propagated rapidly toward the end of the beam.

Failure mode

All the beams subjected to a moving load regardless of span lengths failed in shear, and did not indicate any distress of the reinforcement. For series MA and MB beams, the mode of failure was immediately evident because of the appearance of the dominating inclined crack early in the fatigue lives of the beams. For series MC and MD beams, however, the mode of failure was not immediately evident since the cracks along the length of the beam continued to propagate vertically upward until just before the final few passes when the diagonal crack developed rapidly causing shear failure of the beam. Failure of the beam generally occurred at only one end where the fatal diagonal crack extended through the depth of the beam. Spalling of the concrete and longitudinal cracking also occurred at this side. However, the other end also showed severe damage as the cracks starting from the bottom surface extended almost to the top surface.

Discussion

A remarkable decrease in the fatigue life of the beams under a moving load is obvious from the S–N curves in Fig. 9. For a given magnitude of load, the fatigue life is reduced to about $1/10^4$ of that under a repeated load on a fixed-

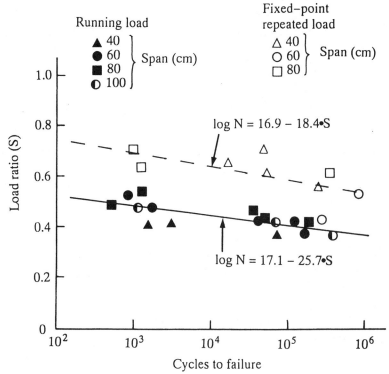

Fig 9   Fatigue strength under a running load and a fixed-point repeated load

point. This reduction could be attributed mainly to the shear stress reversal (Fig. 10) experienced by each point on the beam. This stress reversal caused an effectively larger stress range which is magnified as the load pass over the tip of the crack. Furthermore, the shear stress reversal resulted to the concrete element being subjected to alternating tension and compression stresses. Since concrete is weak in tension, fracture is reached at fewer cycles. Aside from this, as the load moves from one end of the beam to the other end, all the points on the beam are subjected to a combined higher shear and higher moment compared with those under fixed-point repeated load. The effect of this loading

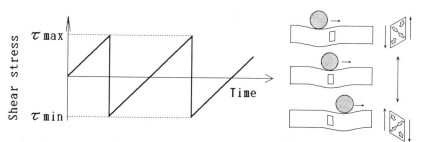

Fig 10(a)   Stress range and stress condition on an element of a beam – moving load

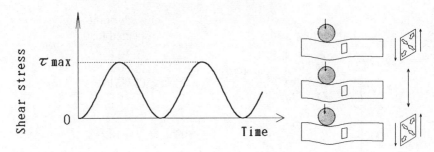

Fig 10(b)   Stress range and stress condition on an element of a beam – fixed point repeated load

condition is manifested in the numerous cracks of the beam which clearly hastened the deterioration of the beam's structural integrity.

Conclusion

The fatigue experiments on microconcrete beams revealed a different behaviour for beams under a moving load as compared with those subjected to fixed-point repeated load. The fatigue strength of a beam decreased remarkably under a running load. This reduction is due to the alternating shear stress induced by moving the load and must be considered in the fatigue design of reinforced concrete bridge beams.

References

(1) KAWAGUCHI, M. *et al.* (1987) Fatigue tests of reinforced concrete slab models of a highway bridge and an attempt to diagnose their residual lives, *Proceedings of the Japan Society of Civil Engineers*, No. 380/I–7, Gihodo Press, Tokyo, pp. 283–292, (in Japanese).

(2) KAWAGUCHI, M. *et al.* (1990) Experiments on the fatigue strength of rc model beams under a running load, Proceedings of JSCE No. 420/V–13, pp. 269–277.

(3) MATSUI, S. and MAEDA, Y. (1986) A rational evaluation method for deterioration of highway bridge decks, Proceedings of JSCE No. 374/I–6, pp. 419–426, (in Japanese).

(4) OKAMURA, H. and HIGAI, T. (1980) Proposed design equation for shear strength of reinforced concrete beams without web reinforcement, Proceedings of JSCE No. 300, pp. 131–141.

(5) SONODA, K. *et al.* (1986) Fatigue failure mechanism of bridge deck RC slabs, Proceedings of the first East Asian conference on structural engineering and construction, (Edited by W. Kanok-Nukulchai, P. Karasudhi, F. Nishino, and D. M. Brotton), Vol. 2, Pergamon Press, pp. 1371–1382.

M. Guagliano* and L. Vergani*

Mixed-Mode Fatigue Crack Propagation in a Crankshaft

REFERENCE Guagliano, M. and Vergani, L., **Mixed-mode fatigue crack propagation in a crankshaft**, *Fatigue Design*, ESIS 16 (Edited by J. Solin, G. Marquis, A. Siljander, and S. Sipilä) 1993, Mechanical Engineering Publications, London, pp. 307–319.

ABSTRACT Experimental fatigue tests were carried out on a mechanical element. Crack propagation was observed to be influenced by all the three fracture modes. Numerical finite element models considered two different crack extension mechanisms. Several calculations were conducted in order to determine: (a) the J integral by means of energy considerations, and (b) the separate stress intensity factors K_I, K_{II}, and K_{III}. An effective stress intensity factor was also determined, utilizing the Tanaka definition.

Notation

a	Crack depth
l	Surface crack length
I, II, III	Modes of fracture
K	Stress instensity factor
k	Normalized stress intensity factor
S	Nominal stress of the uncracked body on the section H–H
N	Number of cycles
C	Tanaka's law coefficient
m	Tanaka's law exponent
J	J integral
E	Young's modulus
v	Poisson's ratio
R_e	Yield stress
R_m	Ultimate tensile strength
n, b, t	Normal, binormal, and tangent vectors

Introduction

The Paris equation (1) successfully correlates the growth rate of long fatigue cracks with the stress intensity factor, K. In the literature it is possible to find expressions for the stress intensity factor for a great variety of situations, most of which are concerned with mode I crack growth due to cyclic tensile loading. In real structural components, however, due to complex geometries and loading conditions, a fatigue crack often propagates under the simultaneous application of classical mode I, mode II, and mode III mechanisms.

In this situation it is very difficult to find exact or closed form solutions to the stress intensity factor. To overcome this problem a number of techniques

* Dipartimento di Meccanica, Politecnico di Milano, Milano, Italy.

have been suggested. In particular the finite element method has been utilized and it is now possible to find a large number of stress intensity factors determined by such a method (2)–(4).

On the other hand other authors have proposed an analytical model for fatigue crack propagation under multiaxial loading. Chen and Keer (5), for example, proposed a direct approach based on a mixed-mode Dugdale model, the accumulated plastic displacement criterion for crack growth and the cyclic J integral, assuming the following:

(1) crack closure and crack branching can be neglected;
(2) the total accumulated plastic displacement is the vector sum of the accumulated crack opening displacement and the crack sliding displacement;
(3) the tension and the shear stress in the yield zone satisfy the von Mises criterion.

The authors developed two equations for small-scale yield conditions; the first is a fourth-power effective stress intensity factor (K_{eff}) crack growth equation, where K_{eff} is defined as

$$K_{eff} = \sqrt[8]{\{(K_I^2 + 3K_{II}^2)^3 \cdot (K_I^2 + K_{II}^2)\}} \tag{1}$$

while the second equation is a second-power cyclic J integral equation.

The authors also made a comparison with experimental results obtained from tests conducted under combined mode I and mode II loading. The results were satisfactory.

Tanaka (6), carried out experimental tests on sheet material with an initial crack oriented at various angles to the longitudinal direction of the plate. He derived a mixed-mode crack propagation law equation considering a Weertman's model and the following assumptions:

(1) the plastic displacements due to cyclic tension and transverse shear are not interactive,
(2) the effective displacement, the accumulation of which results in fracture, is the sum of the displacements of two modes,
(3) the yield stress for shear deformation is one-half of that for tension (criterion of Guest Tresca).

The growth law determined in this way is

$$\frac{da}{dN} = C \cdot (\Delta K_{eff})^m \tag{2}$$

with the effective stress intensity factor

$$K_{eff} = \sqrt[4]{\{(K_I^4 + 8K_{II}^4)\}} \tag{3}$$

and an experimental value of m equal to 4.4.

Tanaka also extended this growth law to the case when the anti-plane shear mode (mode III) cycling is coupled with the above two mode loadings, the effective stress intensity factor, K_{eff}, being

$$K_{eff} = \sqrt[4]{\left\{(K_I^4 + K_{II}^4) + \frac{8K_{III}^4}{(1-v)}\right\}} \qquad (4)$$

but no experimental tests were conducted under this loading condition.

Meguid (7) conducted a three-dimensional computer-aided finite element study to calculate the stress intensity factor K_I for V notched and unnotched circumferentially cracked geometries. In (8) a finite element procedure is used to calculate the stress intensity factors of a weld toe surface flaw with curved crack surfaces in a tubular joint under mixed-mode loading. Other studies on three-dimensional problems could not be found in literature.

In the present paper, experimental fatigue tests have been carried out on a Diesel engine crankshaft, and cracks growing under a three-dimensional stress state have been observed. A three-dimensional numerical model has been realized and structural analyses have been carried out by means of a finite element code and the values of the J integral have also been calculated. The values of the stress intensity factors K_I, K_{II}, and K_{III} along the crack fronts have been calculated using the results of the finite element calculations.

Experimental procedures

In Fig. 1 a section of a marine Diesel engine crankshaft used for the fatigue tests is shown. The crankpin and the journal pin diameters are equal to 130 mm.

Crankshafts were constructed of a steel 35CrMo4 UNI 7874. The mechanical properties determined from cylindrical specimens cut out from a crankshaft are shown in Table 1. In order to load the crankpin with a constant bending moment, an eccentric alternate axial load $\pm \Delta P$ was applied to the crankshaft sections by means of a mechanical device. The distance between the axis of the crankpin and the loading application point is 285 mm.

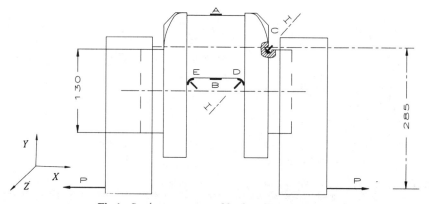

Fig 1   Strain gauge map and load application device

Table 1   35CrMo4 UNI 7874 mechanical characteristics

Yield stress R_e (MPa)	Tensile strength R_m (MPa)	Young's modulus E (MPa)	Poisson's ratio v	Elongation %
572	764	205 000	0.28	9.4

Electrical resistance strain gauges were applied to the crankshaft, in particular at the most stressed zones near the crank fillet. Strain gauge measurements were taken at constant intervals of cyclic loading, the aim being to continuously monitor the axial and circumferential strains. In fact a change in the strain behaviour was generally characteristic of the nucleation of cracks.

After crack nucleation, the fatigue tests are continued and the crack propagation is measured by means of crack propagation gauges (special electrical resistance gauges which consist of a number of resistor strands connected in parallel; when bonded to a structure, progression of a surface crack through the gauge pattern causes successive opening–circuiting of the strands, resulting in an increase in total resistance). The tests are continued until complete rupture of the element has occurred. Different crankshafts have been tested at different load values. In particular the crack evolution observed in the crankshaft loaded by $P = 40\,000$ N is considered in this work.

The fatigue crack begins to grow at the middle point of the internal fillet (see position D, Fig. 1) and propagates in the plane H–H–Z (see Fig. 1) which is inclined at 45 degrees with respect to the load application direction (direction X). The crack assumes a semi-elliptical shape and it is symmetrical with respect to the plane X–Y. However, when the surface crack length l is

Fig 2   Crack surface: the different dimensions of the cracks considered in the numerical models are marked, the fracture plane is at 45 degrees wih respect to the load application direction, except the cross hatched surface, which is inclined at 35 degrees

greater than 60 mm, the plane of propagation changes in the zone near the free surface and it becomes inclined at about 35 degrees with respect to the X axis (see Fig. 1).

Figure 2 shows the cracked section of the crankshaft; it can be seen that the crack front is semi-elliptical and that the crack aspect ratio changes during the propagation. The cross hatched surface indicates the part of the crack surface that lies on a plane at 35 degrees with respect to the X axis (see Fig. 1).

The same characteristics of the fracture surface were observed on the other crankshafts tested in this programme. The crack dimensions shown in the figure (schematized in the numerical models) correspond to a number of cycles equal to $N = 355\,000$ and $N = 540\,000$.

Finite element calculations

Description of the model

Detailed stress intensity factor solutions are necessary to predict fatigue crack propagation and fracture strength of engineering components. Because of the particular geometries of machine elements, it is not always possible to obtain exact solutions. Therefore, it is necessary to use numerical methods or experimental analyses to overcome the problem. The finite element method is now the method which is most widely used to obtain reliable values of the stress intensity factor. The main problem of these types of analyses is that the results are influenced by the mesh adopted. For example the accuracy of the results is strongly dependent on the type of schematization and the elements used. The problem is complicated by the stress singularity that exists at the crack tip which can only be taken into account using particular procedures or elements. During the past years many devices capable of resulting in a good model have been suggested. The common point of the different analyses is that the mesh must be suitably fine near the crack tip if reliable stress and strain values are required. If an energy evaluation of fracture mechanics parameters is conducted, the J integral calculation can be carried out using the stiffness derivative technique: in this case rather coarse meshes are necessary (9).

Due to the computational difficulties, three-dimensional numercal fracture mechanics analyses have only recently been developed. In these cases geometric complications are generally so great that it is necessary to use some solid modelling facility to mesh complicated cracked elements and to check the model.

The crankshaft was modelled utilizing the PATRAN pre-processor in order to obtain a suitable mesh for different crack geometries. The elements were quadratic isoparametric and the zone near the crack tip was schematized by surrounding rings of elements. The use of degenerated elements was avoided to obtain a less complex model. The accuracy of the analyses was increased by the use of the quarter point technique (10)(11) to simulate the elastic crack tip singularity.

Two numerical models were constructed. In both models the lubrication and lightening holes were neglected and the crack was assumed symmetrical with respect to the medium plane of the structure (plane X–Y, see Fig. 1). Consequently, the component becomes double symmetric and it is necessary to schematize only one quarter of the crankshaft; this results in a considerable saving of modelling and calculation times. The constraints imposed were of the symmetrical type and the load application device was also schematized.

An elastic analysis was conducted on every model: the load imposed was of practical significance ($P = 20\,000$ N considering one half of the structure, axial direction, and eccentric position). A convergence study of the results was carried out on the models; when the stress results showed no real change from one model to the finer one, the refinement was stopped. The first model represents the crankshaft with the crack lying on the plane normal to the maximum stress (Fig. 3). A first study on this model has already been carried out considering only the influence of the opening mode (12).

The shape of the crack was assumed semi-elliptical with a crack aspect ratio of 1 : 2, close to the one obtained experimentally. The total surface length l of the crack is 64 mm. The model consists of 796 elements and 4378 nodes, and the total number of degrees of freedom is 13 134.

The second model is similar to the first one, but represents the crankshaft with a crack out of the fillet that is moving towards the web. The change of the plane of the crack has been modelled; particular care was taken to evalu-

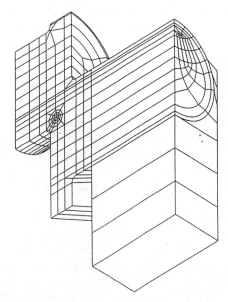

Fig 3   Finite element model of the cracked crankshaft with a surface crack length *l* equal to 64 mm

ate the contributions of the different propagation modes. In this case the shape of the crack is no longer properly semi-elliptical and the crack aspect ratio is approximately 1 : 4: propagation is in fact, faster in the direction of the major axis of the ellipse. The surface crack length l is equal to 120 mm. The model consists of 1844 elements and 9478 nodes, and the total number of degrees of freedom is 28 434.

Stress intensity factor calculations

In order to correctly evaluate the stress intensity factors K_i (i = I, II, III), the three-dimensional stress fields in the region of the crack tip were considered. Using an asymptotic series expansion in three-dimensions Hartranft and Sih (13) showed that on an arbitrarily defined curved crack front the stress field near the crack tip in the planes defined by the normal n and the binormal b (Fig. 4) is the same as the two-dimensional case, even if the values of the stress intensity factors can change along the crack front. This can be written

$$\sigma_n = \frac{K_I(s)}{\sqrt{(2\pi r)}} \cos \frac{\vartheta}{2} \left(1 - \sin \frac{\vartheta}{2} \sin \frac{3\vartheta}{2}\right)$$

$$- \frac{K_{II}(s)}{\sqrt{(2\pi r)}} \sin \frac{\vartheta}{2} \left(2 + \cos \frac{\vartheta}{2} \cos \frac{3\vartheta}{2}\right)$$

$$\sigma_b = \frac{K_I(s)}{\sqrt{2\pi r}} \cos \frac{\vartheta}{2} \left(1 + \sin \frac{\vartheta}{2} \sin \frac{3\vartheta}{2}\right)$$

$$+ \frac{K_{II}(s)}{\sqrt{(2\pi r)}} \sin \frac{\vartheta}{2} \cos \frac{\vartheta}{2} \cos \frac{3\vartheta}{2}$$

$$\sigma_t = 2\nu \left(\frac{K_I(s)}{\sqrt{(2\pi r)}} \cos \frac{\vartheta}{2} - \frac{K_{II}(s)}{\sqrt{(2\pi r)}} \sin \frac{\vartheta}{2}\right)$$

$$\tau_{nb} = \frac{K_I(s)}{\sqrt{(2\pi r)}} \sin \frac{\vartheta}{2} \cos \frac{\vartheta}{2} \cos \frac{3\vartheta}{2}$$

$$+ \frac{K_{II}(s)}{\sqrt{(2\pi r)}} \cos \frac{\vartheta}{2} \left(1 - \sin \frac{\vartheta}{2} \sin \frac{3\vartheta}{2}\right)$$

$$\tau_{nt} = - \frac{K_{III}(s)}{\sqrt{(2\pi r)}} \sin \frac{\vartheta}{2}$$

$$\tau_{bt} = \frac{K_{II}(s)}{\sqrt{(2\pi r)}} \cos \frac{\vartheta}{2} \tag{5}$$

for the stresses with similar formulas for the displacements.

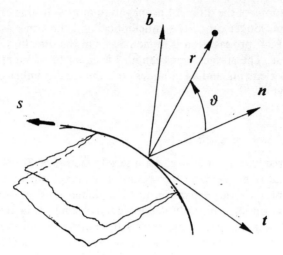

Fig 4   Crack front coordinate system

Considering $\vartheta = 0$, the stress intensity factors can be evaluated by extrapolating the nodal values at distances x to the crack tip

$$K_I = \lim_{x \to 0} \sigma_n \cdot \sqrt{(2\pi x)}$$

$$K_{II} = \lim_{x \to 0} \tau_{nb} \cdot \sqrt{(2\pi x)}$$

$$K_{III} = \lim_{x \to 0} \tau_{bt} \cdot \sqrt{(2\pi/x)} \tag{6}$$

The same extrapolations can be applied to the displacements. More accurate values are generally obtained despite being more sensitive to the angle chosen for the calculations and giving better results for $\vartheta = \pi$ **(14)**. However, in order to obtain reliable results, this type of calculation needs a highly refined mesh and tests to verify the mesh are necessary. Consequently, the J integral was also calculated using a numerical procedure based on Parks' virtual crack extension method. This technique requires rather coarse meshes to obtain good results but does not permit separation of the different contributions of the fracture modes. Bearing in mind that a linear elastic material is being considered, the following relation applies if a plane strain condition is considered

$$J = \frac{(K_I^2 + K_{II}^2)(1 - v^2)}{E} + \frac{K_{III}^2(1 + v)}{E} \tag{7}$$

and a similar one can be used for the plane stress condition. The values of the J integral obtained verify the accuracy of the mesh in terms of stresses and strains and not only for an energy evaluation.

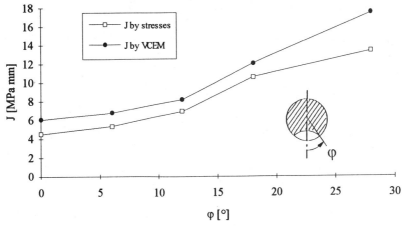

Fig 5 Trend of *J* integral along the crack front (*l* = 64 mm).
● = the values calculated by means of the Virtual Crack Extension Method
□ = the values obtained by means of the numerically determined stresses

Results and discussion

The numerical analyses showed the contributions of the different fracture propagation modes for the two crack lengths examined. The models were first
verified by comparing the strain gauge readings with finite element strains
(points A–E, Fig. 1); the deviations were less than 10 percent. The values of the
stress intensity factors K_i (i = I, II, III) were evaluated by extrapolating (to the
crack tip) the nodal values obtained from the stresses given by equation (6).
This type of calculation is not, generally very accurate, but a validation was
found by comparing the *J* integral values from equation (7) with those

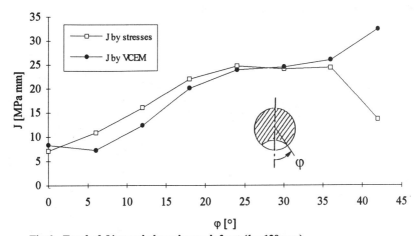

Fig 6 Trend of *J* integral along the crack front (*l* = 120 mm).
● = the values calculated by the Virtual Crack Extension Method
□ = the values obtained by means of the numerically determined stresses

obtained using the virtual crack extension method. The trend of the J integral along the crack front for the two cracks considered is shown in Figs 5 and 6. In both cases the agreement between the two types of calculation is good, except for the points at the free surfaces. In Fig. 7 the trend of the normalized stress intensity factors ($k_i = K_i/S_{ij}\sqrt{(\pi a)}$, where $ij = nn$, nb, and a is the crack depth) is shown for the smallest crack considered ($l = 64$ mm). The maximum values of k_i are not at the deepest point of the crack; this is probably due to relaxation of the back surface of the crankshaft and to the effects of the tri-axiality of the stress state. For simplicity the section used for the calculation of S is considered with constant thickness along the z direction.

The values of the normalized k_{II} stress intensity factor are greater than those of k_I: however its influence on the propagation is minor because of the smaller nominal stress.

In Fig. 8 the behaviour of the stress intensity factors I, II, and III are shown for the longest crack modelled. In this case the third mode of propagation is also present. The values are all normalized. It is noted that the influence of the second and the third mode becomes stronger when the plane of propagation changes. It is necessary however, to remember that the values of the nominal tangential stress are numerically smaller than the nominal tensile stress. The effective stress intensity factor K_{eff} is also calculated using the expression of Tanaka, see equation (4).

The experimental tests carried out permitted researchers to follow the crack propagation on the surface. In Fig. 9 the trend of the surface crack length is shown as a function of the cycle number and it can be assumed that the stress intensity factor variations on the surface occur without a strong discontinuity in the crack growth even if the plane of propagation changes. This can be explained by looking at the effective stress intensity factor, calculated according to the Tanaka definition given by equation (4).

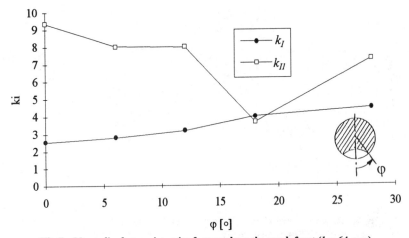

Fig 7   Normalized stress intensity factors along the crack front ($l = 64$ mm)

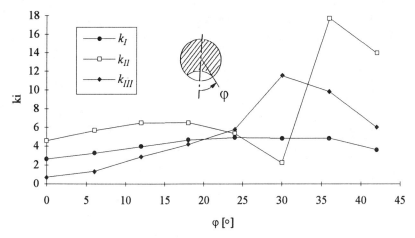

Fig 8   Normalized stress intensity factors along the crack front ($l = 120$ mm)

In Fig. 10 the non-dimensional ($k_{eff} = K_{eff}/S\sqrt{\pi a}$) trend for the two crack configurations are shown. It can be seen that in both cases the normalized effective stress intensity factors on the free surface are similar to k_I; this is probably why the sudden change in crack growth is not observed. In the inner part of the crack front, however, the propagation is governed not only by the opening mode but also by the other modes. In both the analyses the greatest values of the effective stress intensity factor are towards the external surface; this explains the faster propagation of the crack along the major axis of the ellipse, which is approximately the shape of the crack front.

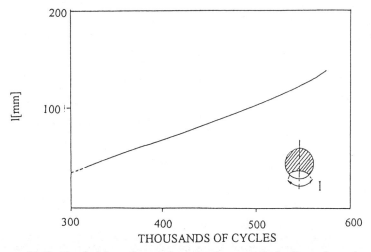

Fig 9   Trend of the surface crack propagation versus the loading cycle number

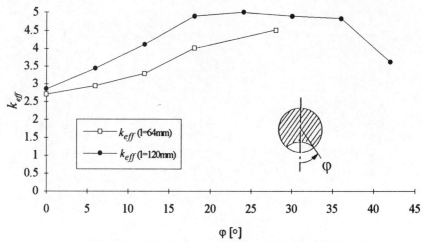

Fig 10   Trend of the normalized effective stress intensity
● = the k_{eff} pattern for the long crack ($l = 120$ mm)
□ = the k_{eff} pattern for the small crack ($1 = 64$ mm)

Conclusions

Fatigue tests were conducted on a crankshaft to study crack propagation under three-dimensional stress state. Numerical analyses carried out by means of a finite element code allowed calculations of the J integral and the values of the stress intensity factors K_I, K_{II}, and K_{III} to be obtained. Good agreement was obtained and it was possible to point out the different contributions made by the three propagation modes. At the external surface the dominant mode was K_I, while at the inside crack front the other modes became larger. An accurate crack growth prediction for mechanical elements of complex geometry must consider all the fracture modes. An effective stress intensity factor was determined using the Tanaka formulation.

Acknowledgement

This work was supported by the Italian National Committee for Researches grant number C.T. 91.03068.CT07.

References

(1) PARIS, P. C. and ERDOGAN, F. (1963) A critical analysis of crack propagation laws, *J. bas. Engng*, 528–533.
(2) FAWKES, A. J., OWEN, D. R., and LUXMOORE, A. R. (1979) An assessment of crack tip singularity models for use with isoparametric elements, *Engng Fracture Mech.*, **11**, 143–159.
(3) WU, X. R. (1984) Stress intensity factors for half-elliptical surface cracks subjected to complex face loadings, *Engng Fracture Mech.*, **19**, 387–405.

(4) NEWMAN, J. C. and RAJU, I. S. (1981) An empirical stress-intensity factor equation for surface crack, *Engng Fracture Mech.*, **15**, 185–192.
(5) CHEN, W-R. and KEER, L. M. (1991) Fatigue crack growth in mixed mode loading, *J. Engng Mater. Technol.*, **113**, 222–227.
(6) TANAKA, K. (1974) Fatigue crack propagation from a crack inclined to the cyclic tensile axis, *Engng Fracture Mech.*, **6**, 493–507.
(7) MEGUID, S. A. (1990) Three-dimensional finite element analysis of circumferentially-cracked notched and un-notched cylindrical components, *Engng Fracture Mech.*, **37**, 361–371.
(8) CHONG RHEE, H. and SALAMA, M. M. (1987) Mixed mode stress intensity factor solutions of a warped surface flaw by three dimensional finite element analysis, *Engng Fracture Mech.*, **28**, 203–209.
(9) PARKS, D. (1979) A stiffness derivative finite element technique for determination of elastic crack tip stress intensity factors, *Int. J. Fracture*, **10**, 487–502.
(10) BARSOUM, R. S. (1976) On the use of isoparametric finite elements in linear fracture mechanics, *Int. J. numer. Methods Engng*, **10**, 25–37.
(11) HENSHELL, R. D. and SHAW, K. G. (1975) Crack tip elements are unnecessary, *Int. J. numer. Methods Engng*, **9**, 495–507.
(12) GUAGLIANO, M. and VERGANI, L. (1991) Propagazione di cricche per fatica in stato di sforzo triassiale, *IGF7, Atti del VII Convegno Nazionale Gruppo Italiano Frattura*, pp. 139–147.
(13) HARTRANFT, R. J. and SIH, G. C. (1977) Stress singularity for a crack with an arbitrarily curved front, *Engng Fracture Mech.*, **9**, 705–718.
(14) MEGUID, S. A. (1989) *Engineering fracture mechanics.* Elsevier Science, Oxford, pp. 151–158.

A. Siljander, M. Lehtonen*, G. Marquis,* J. Solin,* J. Vuorio,†
and P. Tuononen†*

Fatigue Assessment of a Cast Component for a Timber Crane

REFERENCE Siljander, A., Lehtonen, M., Marquis, G., Solin, J., Vuorio, J., and Tuononen, P., **Fatigue assessment of a cast component for a timber crane,** *Fatigue Design,* ESIS 16 (Edited by J. Solin, G. Marquis, A. Siljander, and S. Sipilä) 1993, Mechanical Engineering Publications, London, pp. 321–331.

ABSTRACT This paper describes the fatigue life assessment of a cast slewing system base which is one of the primary load carrying components in log loaders. Field measurements were conducted and strain histograms were produced to quantify the required fatigue strength for the next generation of components made of a new material and with larger payload capacities. Finite element analyses of the cast base subjected to the measured external loads provided vital information on the stress and strain distributions for subsequent fatigue life assessment and allowed alternate designs to be simulated rapidly. Recommendations on the geometry of the new component could be made early in the product design cycle when these design-integrated approaches were incorporated.

Introduction

Background

Design-integrated mechanical testing and analyses are being highlighted via various case studies in an on-going national project 'Weldable high strength steels', coordinated by the Federation of Finnish Metal Engineering and Electrotechnical Industries (FIMET). A primary objective of this project is to assist engineers in utilizing high strength cast steels in design applications where reduced weight and high fatigue strength are of primary concern.

This paper briefly describes one of the case studies within the project concerning the re-design of a log loader slewing system base using a high strength cast steel. The study was performed in the autumn of 1991 and aimed to quantify the design requirements of a cast slewing system base commonly employed in log loaders.

Overview of the slewing system base

Figure 1 shows a FMG 910 forest tractor, manufactured by FMG Timberjack, equipped with a hydraulic Loglift F 60 FT 103 timber crane, manufactured by Loglift Oy Ab. Forest tractors of this type are widely used to collect cut timber

* Metals Laboratory, Technical Research Centre of Finland (VTT), P.O. Box 26 (Keministinie 3), SF-02152 Espoo, Finland.
† Ship Laboratory, Technical Research Centre of Finland (VTT), P.O. Box 26 (Keministinie 3), SF-02152 Espoo, Finland.

Fig 1   The FMG 910 forest tractor in use

from those rough-terrain areas that are unaccessible by conventional timber trucks. Logs are loaded onto the cargo area of the tractor and transported to roadsides for later collection by timber trucks, which then haul the cargo to various woodpulp manufacturing plants in Finland.

The cast slewing system base is one of the primary load carrying components of the timber crane. Because its fatigue strength is of primary concern for safety reasons, the slewing system base was selected as the object of this study. The base is a complex casting which sustains a variety of operational loads. These include

(a) a bending moment, M, resulting from the offset timber lifting operation;
(b) a downward vertical force, N, due to the timber weight; and
(c) horizontal forces, F_s, applied through the rotational actuator mounting bolts.

The bending moment, M, is resolved into two horizontal forces, P_M, acting through the upper and lower bearings. The vertical force, N, is assumed to be fully carried by the lower bearing. A schematic of the assumed forces is presented in Fig. 2. The base is secured with bolts to the steel frame of the tractor.

Aim and scope

The ultimate goal is to redesign the component for future models of forest tractors having larger payload capacities without increasing component weight or manufacturing costs and while maintaining interchangeability with

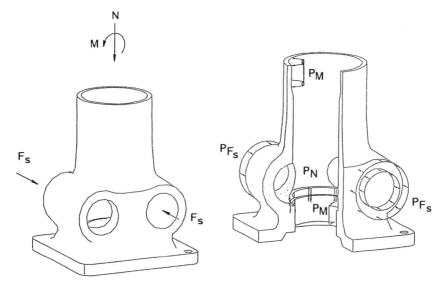

Fig 2   Schematic of the slewing system cast base and the assumed operational load components

existing slewing system bases. Because an increased payload causes higher operating stresses in the fatigue critical locations, a new high strength high toughness cast steel was recommended. To ensure the adequate fatigue strength of the increased capacity components, an integrated design strategy involving both experiments and analysis was implemented.

Within the framework of the above requirements, field measurements were conducted with the existing cast component and operational duty cycle data for subsequent fatigue analysis were collected. Finite Element (FE) analyses and fatigue life predictions were then performed to determine the objective design specifications that should be used in redesigning the log loader's slewing system base. This paper outlines the experimental and analytical methods employed in the design-integrated fatigue life assessment.

Experimental program

Field measurements

Due to the lack of operational loads data for the cast base, field measurements were carried out to capture operational duty cycle data for subsequent event identification, stress analysis, and fatigue life estimation (1)(5)(6)(9). Strain gauges were fixed to selected locations of the slewing system base. Strain signals were recorded with an instrument recorder for later analyses. All measurement hardware was installed in the operator cab of the tractor as depicted in Fig. 3.

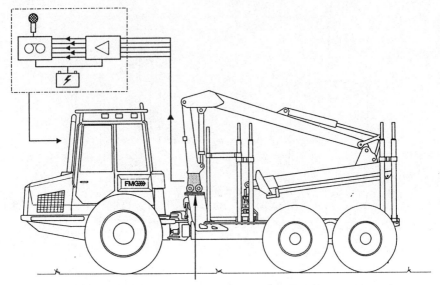

Fig 3 Overview of the field measurement set-up. All strain gauges were mounted on the base of the slewing system (shaded area)

Field measurements of characteristic operational load events were collected during normal operation on a typical rough-terrain harvesting area. Each duty cycle consisted of the following activities:

(a) manoeuvring the vehicle into close proximity of a cut timber pile;
(b) loading the timber onto the cargo area using the tractor's own hydraulic crane;
(c) manoeuvring to a new location, loading etc. until the cargo area was fully loaded;
(d) transporting the payload to a road accessible by timber trucks, and finally;
(e) unloading the timber, again using the tractor's own hydraulic lift.

The field measurements involved collecting strain gauge data during a series of typical duty cycles. Based on the measured variable amplitude strain histories, two-dimensional Rainflow histograms were formed. The strain histograms could be used for direct fatigue life prediction for the instrumented details based on the material's S–N-curve. Figure 4 shows a representative two-dimensional Rainflow histogram collected from one of the strain gauges during the field measurements.

Some of the variable amplitude strain signals were converted to variable amplitude external loads to conduct fatigue life predictions based on the entire FE-model.

In addition to the variable amplitude duty cycle documentation, numerous static timber lifting tests were recorded. The static tests involved lifting timber of known weight from various directions and distances with respect to the slew

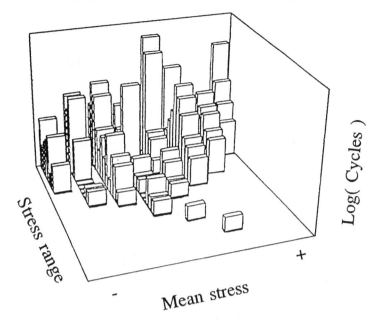

Fig 4   A two-dimensional Rainflow table constructed from strain measurements at one location on the slew system base during a series of duty cycles

system base. The most critical loading maxima and minima were identified from the static and dynamic field measurement strain time-histories. The extreme strain values were converted into loads to be used in the subsequent FE-analyses.

Materials data

OS 690 Vaculok was recommended as a suitable material for the next generation slew system base. The OS Vaculok class of steels have been developed at Rauma Oy Materials Technology and combine high yield strength with excellent fracture toughness (2)(4). Fatigue data for this material already existed (3) and are shown here in Fig. 5. These data were employed in the fatigue life estimations. Material properties for this steel are shown in Table 1.

Analysis and results

Finite element analysis

A three-dimensional FE-model of the entire slewing system base was created using PATRAN pre-processor (7) running on a 16 Mb Silicon Graphics 4D/25G workstation. Eight node isoparametric elements were used throughout the model. The coarse element mesh in the upper part of the casting was used to transfer the loads to the more interesting regions of the base, where a

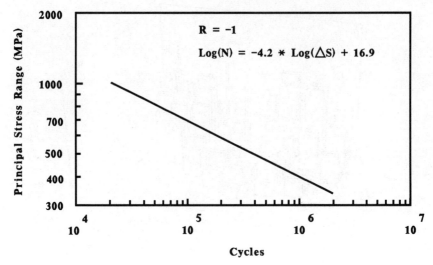

Fig 5   S–N curve for OS 690 Vaculok cast steel (3)

finer element mesh was employed. Figure 6 shows the three-dimensional FE mesh. In the final FE-model, there were 4460 nodes, 3013 elements, and 13 744 degrees of freedom.

The FE-model was first loaded statically such that the timber lifting orientation angle, lifting distance, and timber weight corresponded to the initial static field measurement conditions described above. The FE-model was then verified by confirming satisfactory correlation between the measured and calculated strains for the given load case. The ADINA structural analysis code was used for the analyses together with the material properties for OS 690 Vaculok cast steel. Due to the large model size and the non-linear boundary conditions, approximately 12 h of CPU time was required to solve one loading case.

Having verified the FE-model with the static loading cases and by knowing the measured strains as a function of the external loads, variable amplitude strain histories were converted to operational load histories. In order to focus on the main goal of the study, i.e., to account for the increased payload capacity of the next generation of timber cranes, these variable amplitude load histories were scaled up so as to represent the larger external loads anticipated in future components. Finally, these 'future design' variable amplitude loading histories were employed in connection with the FEA results to produce variable amplitude stress time-histories which were then used for fatigue life prediction.

Table 1   Typical mechanical properties of the OS 690 Vaculok cast steel
(R_m = UTS, R_{el} = σ_{ys}) [3]

Material	R_m (MPa)	R_{eL} (MPa)	A_5 (%)	Z_5 (%)	Charpy 'V' $J/°C$
OS 690 Vaculok	1010	930	17	63	85/-40

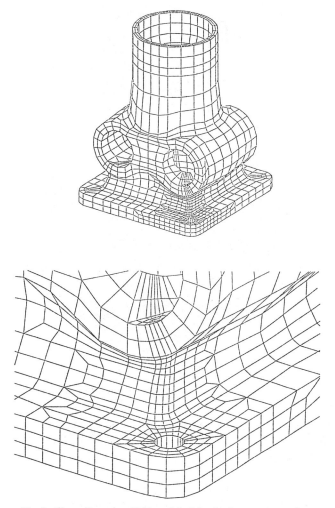

Fig 6   Three-dimensional FE-model of the slewing system cast base

Fatigue life predictions

Fatigue life predictions were conducted on the entire three-dimensional FE-model of the slewing system base. The stress analysis was linearly elastic so all stress and strain components in the FE-model were proportional to the applied loads. Patran's fatigue life evaluation package, P/FATIGUE, was one of the post-processors employed in the fatigue life assessments. The nominal stress–life approach using the local principal stresses from the FE-analysis with Goodman's mean stress correction and the S–N curve for OS 690 Vaculok were utilized. An example of a post-processor fatigue life contour plot for the cast model is presented in Fig. 7.

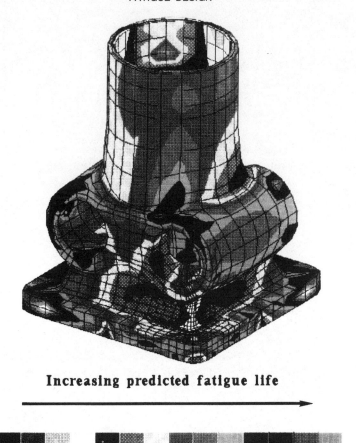

Increasing predicted fatigue life

Fig 7   An example plot of the estimated fatigue life contours of the initial geometry

By using a three-dimensional FE-model and a post-processor capable of fatigue life assessment, like P/FATIGUE, an efficient system for exploring possible design modifications exists. Based on the initial fatigue life estimations of the three-dimensional FE-model, some structural changes were implemented in the FE-model. The purpose of these structural modifications was to study the influence of geometry alterations on the predicted fatigue life. Within the limits set by the casting process, some of these structural changes were implemented in the design of the slewing system cast base.

Discussion

Stress histories

Strain signals at multiple locations on the timber crane base were recorded during a series of typical duty cycles of a forest tractor. The fatigue life predictions of the slewing system base were conducted using the measured service histories. Therefore, it is believed that the analyses described above give a more realistic life prediction than if the operating loads had been rough-guessed. In anticipation of a new generation of higher capacity log lifters, the resulting extrapolated operational loads were then scaled and used in the FE analyses for stress–strain determination and ultimately for component geometry alteration.

This case study has been limited to the cast slew system bases used in timber cranes mounted on forest tractors. Similar cast components are also used for log lifters installed in process environments, e.g., at woodpulp plants, and for the hydraulic cranes of timber trucks. In process environments, cranes are typically mounted on more rigid foundations and lift only from one direction and distance and unload at another, well defined location. Therefore, the magnitude and distribution of operational stresses are expected to be totally different than for forest tractor mounted units. Similar statements also apply to the slew system bases of log lifters mounted on timber trucks. The methods outlined in this example illustrate a logical design strategy for ensuring fatigue strength which may be expanded to include other applications of slew system bases.

Design strategy

Figure 8 schematically shows the various steps and procedures which should be followed in an overall design strategy for reliable fatigue resistant components and structures (8). These steps can be enumerated as:

(a) the expected service loading spectrum is predicted based on actual field measurements, standardized histories, or engineering estimates;
(b) elastic stress concentration factors at the fatigue critical locations are determined using FE-analysis, handbook approximation or another suitable method;
(c) cyclic strength and fatigue properties of the candidate material(s) are measured;
(d) the nominal spectrum stresses are converted to give the localized stress–strain behaviour at the critical notches;
(e) expected fatigue life and reliability are estimated;
(f) the proposed design can be modified to improve fatigue resistance or optimize component weight and shape (steps (b)–(f) can be repeated until a satisfactory design is achieved); and
(g) spectrum fatigue testing of prototypes or final components should be included to verify the design and, if needed, lead to further refinements.

With respect to the fatigue life assessment of the cast slewing system base detailed in this paper, the steps (a)–(f) above have already been completed within this ongoing project. In step (a), operational loads have been determined for the most important application of the case bases. However, longer

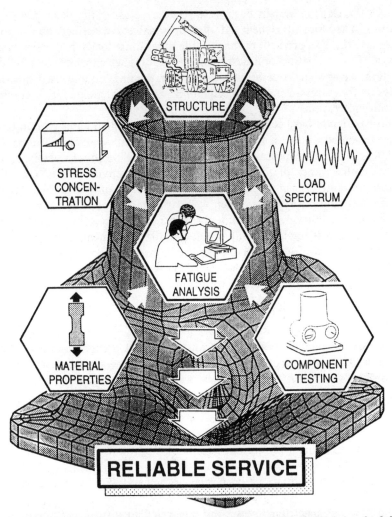

Fig 8 Schematic illustrating the overall strategy used in the design of reliable fatigue loaded components

term measurements and data from other potential usage situations will be useful for making final fatigue reliability estimates. Spectrum fatigue testing of prototypes can ultimately help ensure component performance.

Conclusions

This paper has outlined the steps taken for making fatigue life assessments and design modification recommendations for a cast slewing system base, a critical load bearing component of the hydraulic log loaders typically used on rough-terrain forest tractors. These steps constitute important elements of an overall strategy for the design of fatigue resistant components. Conclusions from this case study can be summarized.

(1) Field measurements from the actual working environment of the forest tractor provided the input data, i.e., magnitude and distribution of operating stresses, needed to realistically estimate fatigue life.

(2) Mechanical material testing data combined with FE-analyses proved instrumental in the objective evaluation of the current geometry and led to recommendations for improving the fatigue strength of the forest tractor's cast slewing system base while staying within the defined restraints for geometry, cost, weight, and interchangeability.

(3) The results of this study will be implemented in the next generation of cast slewing system base components.

(4) The final verification of the fatigue life estimation of the slewing system base involving spectrum fatigue testing of prototypes remains to be done.

Acknowledgements

The work reported here is part of an on-going national project 'Weldable high strength steels', coordinated by the Federation of Finnish Metal Engineering and Electrotechnical Industries (FIMET). The authors would like to acknowledge Dr J. Liimatainen of Rauma Oy Materials Technology, Mr I. Inkinen of Loglift Oy Ab, and FMG Timberjack for their cooperation and for their kind permission to publish this work.

References

(1) BUXBAUM, O. (1979) Random load analysis as a link between operational stress measurement and fatigue life assessment, *Service fatigue loads monitoring, simulation, and analysis, ASTM STP 671*, (Edited by P. R. Abelkis and J. M. Potter), ASTM, Philadelphia, pp. 5–20.

(2) LIIMATAINEN, J. (1990) *Extra high strength low carbon Ni-Cr-Mo cast steel*, PhD Thesis. Tampere University of Technology, Finland.

(3) LIIMATAINEN, J. (1991) Rauma Oy Materials Technology, Tampere, Finland, private communication.

(4) MARTIKAINEN, H. O. (1987) New developments in cast arctic offshore steels, Proceedings of the 19th Offshore Technology Conference, pp. 15–19.

(5) MARTZ, J. W., SMILEY, R. G., and KORMOS, J. G. (1978) Field testing of reference vehicles as an aid to the design analysis process for earthmoving equipment, *SAE paper No. 780 485, Earthmoving Industry Conference*, Society of Automotive Engineers.

(6) Design data base aids product development, *SDRC Newsletter*, **12**, Structural Dynamics Research Corporation pp. 3–9.

(7) PATRAN version 2.5, User's Manual. PDA Engineering, 1991.

(8) SOLIN, J. (1990) Application of local strain approach for steels OS 540 and Polarit 778, *Fatigue Design 1990*. (Edited by J. Solin (Technical Research Centre of Finland, Espoo, pp. 33–93.

(9) SOLOMON, A. and BRANCH, R. (1991) Mechanical testing in design, *Machine Design*, 110–118.

*E. Niemi**

Aspects of Good Design Practice for Fatigue-Loaded Welded Components

REFERENCE Niemi, E., **Aspects of good design practice for fatigue-loaded welded components,** *Fatigue Design*, ESIS 16 (Edited by J. Solin, G. Marquis, A. Siljander, and S. Sipilä) 1993, Mechanical Engineering Publications, London, pp. 333–351.

ABSTRACT The importance of understanding the various sources of stress concentrations is discussed. The structural discontinuities are classified into three categories and the significance of this distinction on the fatigue analysis is explained. Various methods of solving structural stresses at the hot spots are discussed. The various ways to improve the design are explained in the light of fracture mechanics. Finally, some recommendations are given for the treatment of variable amplitude loading.

Introduction

In the design of welded structures, fatigue is often a failure mode which is overlooked. Serious fatigue failures in the offshore branch have led to huge research projects and to a large number of international conferences on the subject. Consequently, the designers of offshore structures are now well aware of fatigue failures and the design practice necessary to avoid them.

Another branch where fatigue of welded components usually represents the governing failure mode is the design of mobile equipment, such as cranes, excavators, and forest harvesters. In the design practice of such structures, the functional requirements play the primary role. Should fatigue cracking occur, it would not seriously endanger the safety of people. Good design practice which considers fatigue serves industry's general quality policy. In this branch, fatigue damage assessments have become increasingly important during the last decade, since designers have attempted to save weight by using newer steels of higher strength.

Traditionally, attempts to counter fatigue cracking have been based on trial and error. The designers seldom possess a deep understanding of the real causes of cracking. Minor changes in detail design are more or less based on intuition and experience. The weld detail classifications of fatigue design codes (1)–(3) have provided some assistance to designers for comparing the relative merits of some basic types of welded details, Fig. 1. The problem with such an approach is that the real details of mobile equipment seldom correspond to the categorized details based on relatively simple test specimens and simple loading conditions.

The increased understanding of the fatigue crack propagation at welds, based on linear-elastic fracture mechanics, and the possibility to analyse welded components using the finite element method, FEM, offer new

* Department of Mechanical Engineering, Lappeenranta University of Technology, Lappeenranta, Finland.

Type/ No	Joint configuration showing mode of fatigue cracking and stress considered/Schéma de l'assemblage indiquant le mode de fissuration	Class/ Classe
17		63
18		71
19		50

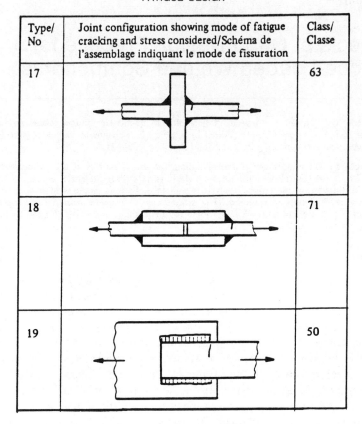

Fig 1   An example of joint classification for alternative detail designs (1)

opportunities to raise the design practice to a new level. The book written by
Gurney (4) has been an excellent reference in this respect. In this paper, some
aspects of such design practice are discussed, focusing primarily on mobile
equipment.

Stress concentrations

It is very important to recognize the ways in which geometric discontinuities
in a structure generate stress concentrations. Moreover, it is useful to divide
the details into three categories according to their global or local nature, as
explained in the following sections (5)(6).

Illustrative example

Figure 2 shows an example which includes two brackets for connecting a
hydraulic actuator, welded on the upper flange of a beam with a box-type
cross-section. In such an example, several stress-raising phenomena can be
identified. By using FEM, and modelling the detail in various ways, it has been

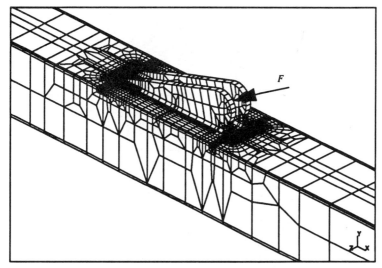

Fig 2   An example of a box-shaped beam with a pair of brackets welded on the flange

possible to differentiate between the following effects causing stress concentration in the flange at the end of the bracket:

(1) concentration of the membrane stress due to the cross-sectional area of the
    bracket;
(2) plate bending stresses due to the eccentricity of the bracket;
(3) plate bending stresss due to the bending stiffness of the bracket which does
    not allow the flange to follow the curvature of the beam as a whole;
(4) plate bending stresses due to the loads acting on the bracket.

Surprisingly, all the above-mentioned effects in this example are roughly of
equal significance. Figure 3 shows a deformed structure and Fig. 4 shows the

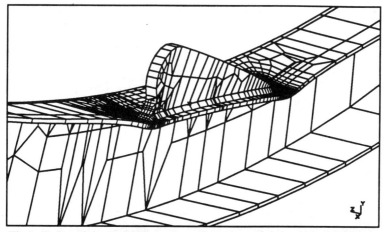

Fig 3   Deformed shape of the flange due to the various effects caused by the brackets

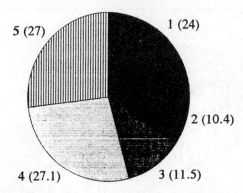

Fig 4 Structural hot spot stress in the flange at the end of the bracket.
   1. Nominal stress
   2. Membrane stress concentration
   3. Plate bending stress caused by restricted beam bending
   4. Plate bending stress due to the bracket eccentricity
   5. Plate bending stress due to bracket loads

relative contributions to the stress increase from the above-mentioned sources, (1)–(4). This kind of study gives the designer a good insight into the real sources of the stress concentration and it also makes a realistic comparison of the design alternatives possible.

Three levels of stress concentration

Depending on the class of stress concentrations which are included in the calculated stresses, the fatigue analysis of welded components can be based on: (a) nominal stresses, (b) hot spot stresses; or (c) notch stresses (or strains) at local discontinuities, including non-linear stress peaks.

The nominal stresses are solved assuming elastic behaviour and using the usual formulae from elementary text books. However, there are some geometric discontinuities of a macro-geometric nature, as well as local areas of con-

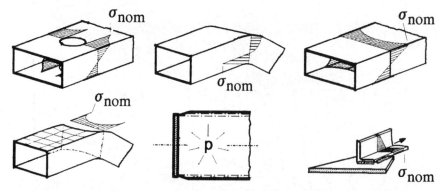

Fig 5 Examples of macro-geometric discontinuities to be taken into account in the determination of nominal stresses

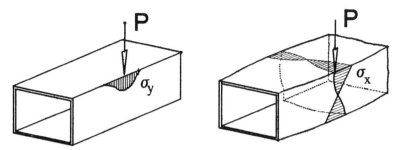

Fig 6 Examples of nominal stresses caused by concentrated loads

centrated loading, which cause global changes in the stress distribution, see Figs 5 and 6. Such effects are not covered by the discontinuities in the test pieces from which the S–N curves have been derived in the fatigue design codes. Therefore, such macro-geometric stress concentrations must be taken into account when nominal stresses are calculated.

Structural stresses in welded structures are compatible with the stress definitions used in the shell theory, see Fig. 7. Membrane stress is the average stress across the plate thickness, and bending stress is the antimetric stress distributed linearly across the plate thickness. Hot spot stress is the value of the structural stress, i.e., sum of the membrane and bending stresses, on the surface at a critical point, called a hot spot, where fatigue cracking is expected. When hot spot stresses are calculated, the structural discontinuities of a more local nature are also taken into account, see Fig. 8. Such structural discontinuities have been present in the test specimens on which the general design codes **(1)–(3)** are based. Therefore, their effects are excluded from the nominal stress. However, whenever the analysis is based on a hot spot S–N curve, on fracture mechanics, or on notch stress/strain methods, the effects of local structural discontinuities must be included in the calculated hot spot stress.

Local notches (see Fig. 9) cause non-linear stress peak distributions which are superimposed on the structural stress, see Fig. 10. The non-linear stress peak is the non-linear part of the total notch stress distribution, and it is in equilibrium. In the literature, different meanings of the term peak stress exist. Therefore, an unambiguous definition has been introduced in reference **(5)** – the non-linear stress peak; its distribution is self-equilibrating.

$$\sigma_s \;=\; \sigma_m \;+\; \sigma_b$$

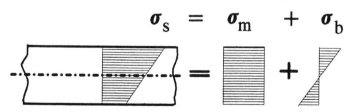

Fig 7 Structural stress consisting of membrane stress and bending stress components

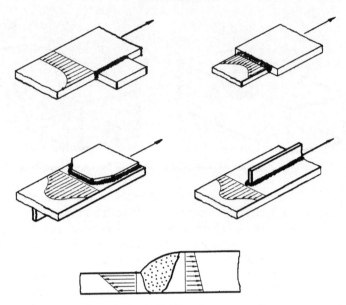

Fig 8   Examples of structural discontinuities which have been present in numerous test pieces used for fatigue testing

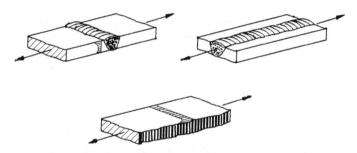

Fig 9   Examples of local notches in welded components causing non-linear stress peaks

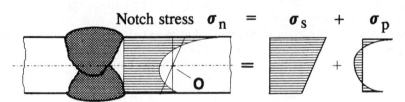

Notch stress σ_n = σ_s + σ_p

Fig 10   Distribution of the total notch stress at a notch consisting of structural stress and non-linear peak

Figure 11 summarizes how the various stress raising effects, stress categories, and fatigue analysis approaches (except fracture mechanics) are interconnected.

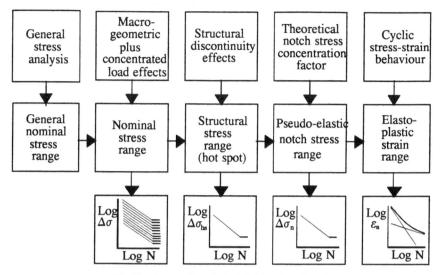

Fig 11   An overview of the definitions introduced

Choice of the fatigue analysis method

The nominal stress method, based on a set of S–N curves is suitable for use when:

(1) the detail corresponds well to one of the classified details;
(2) the nominal stress can easily be calculated;
(3) the fabrication tolerances meet the requirements set for misalignments.

The hot spot method, based on structural hot spot stresses and on a corresponding S–N curve, is to be preferred when:

(1) strain range recordings are made at hot spots during prototype test runs;
(2) the finite element method is used for stress determination using shell or solid elements;
(3) the expected crack initiation occurs at a toe of a transverse weld or at the end of a weld;
(4) the structural discontinuity is not comparable with any classified details included in the design rules (nominal stress approach).

The fracture mechanics approach is a very versatile method, particularly when a damage-tolerant design is desired, or fitness for purpose of a structure containing excessive discontinuities should be assessed. Fracture mechanics analysis of crack growth (see Fig. 12) yields information about:

(1) the expected life of a welded joint;
(2) the remaining life of a cracked part;
(3) the tolerable crack (discontinuity) size;
(4) the required fracture toughness of the material;

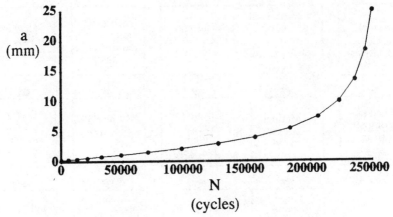

Fig 12   An example of propagation of a semi-elliptic crack through the plate at a weld toe

(5) the frequency of the in-service inspections required;
(6) the required accuracy of the in-service inspections; and
(7) the effects of proposed improvements in design or fabrication.

The original form of the notch stress/strain method, often called the strain–life approach (7) is based on notch strains determined by using the cyclic stress/strain curve and the Neuber's rule. This approach is most suitable for the analysis of low cycle fatigue, i.e., when cyclic plasticity occurs at a notch. The simpler notch stress methods are suitable for studying the effects of the shape of the weld reinforcement and the effect of improving the shape in the high cycle regime. Some newer versions of the notch stress method have proved suitable for cases with non-proportional multiaxial fatigue of welded joints (8).

Determination of the hot spot stress

Three basic methods can be used for the determination of structural stresses at structural discontinuities:

(1) measurement of the strains from models or prototypes, usually using strain gauges;
(2) finite element analysis of the structural component using shell or solid element modelling;
(3) multiplication of the nominal stresses by a structural stress concentration factor, K_s.

The location and orientation of the strain gauges should be chosen carefully in order to find the real hot spots. If in doubt, prior use of a brittle lacquer is a good alternative. At least two strain gauges, suitably placed (5), are needed because the results must be extrapolated to the weld toe, Fig. 13. Some strain gauges should be placed away from the discontinuity in order to determine the nominal stresses.

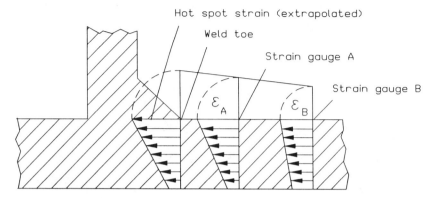

Fig 13   Linear extrapolation of the structural stresses or strains to the weld toe

It is not an easy task to perform finite element analyses of welded structures in order to resolve structural hot spot stresses. The analyst must have experience with such work and have a good insight into the goals of the analysis. One of the difficulties is the choice of the correct element type and size. One typical pitfall is shown in Fig. 14. It is easy to model a longitudinal stiffener with thin shell elements. However, the interpretation of the results in the plate at the end of the stiffener is quite difficult. In reality, the structure looks like that modelled with solid elements in Fig. 15. Thin shell elements are easy to use but they do not yield correct results in the vicinity of the hot spot **(9)**. Solid elements yield better results but are more tedious to use. An experienced analyst can interpret the thin shell results in such a way that the results are satisfactory.

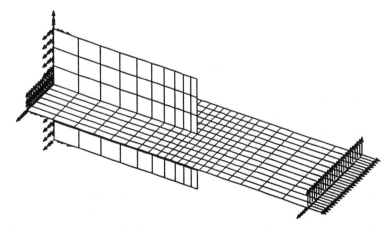

Fig 14   An example of a thin shelled model which does not directly yield correct structural stresses in the plate at the end of the gusset

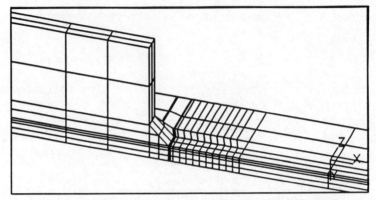

Fig 15   An example of solid element modelling of a gusset plate detail

The use of relevant stress concentration factors, K_s, would be the easiest way to solve structural hot spot stresses. Unfortunately, there is a lack of such factors in the form of parametric formulae for most structural details. One of the few types of detail which is well documented, is the welded joint between circular hollow sections **(10)**. Recently, some work has been carried out concerning ship hull details, for example **(11)**. It is most important that authors produce factors that are determined in a consistent way. A proposal for a harmonized practice is presented in reference **(5)**.

Methods of improving the design

Factors affecting the fatigue life

Fatigue crack propagation life can be expressed by the following formula which is derived by integrating the well-known Paris' relation

$$N_f = \frac{1}{\pi^{1.5} C(R) K_s^3 \Delta\sigma^3} \int_{a_0}^{a_f} \frac{da}{F^3(a) M_k^3(a) a^{1.5}} \tag{1}$$

where

N_f is the number of cycles to final crack size;
a_0 is the size of the initial crack;
a_f is the size of the final crack;
$C(R)$ is a constant depending on the stress intensity factor ratio, R;
K_s is the structural stress concentration factor;
$\Delta\sigma$ is the nominal stress range;
$F(a)$ is a factor depending on the shape and size of the crack;
$M_k(a)$ is a factor depending on the weld reinforcement and the crack size;
a is the size of the crack at a particular time.

Equation (1) can be programmed easily using a microcomputer such that all the usual welded details can be studied. Figure 12 shows an example of such a

study. The functions $F(a)$ and $M_k(a)$ are described for example in reference
(12). Moreover, it helps us to make qualitative conclusions about the various
ways of improving the design. Such improvement possibilities are discussed
below.

Reduction of the nominal stress range

The most trivial way to improve fatigue life is to reduce $\Delta\sigma$ by choosing a low
strength material and designing a heavy structure. However, considerably
better methods are available. The main rule is not to place welds with low
fatigue strength in areas containing high nominal stresses, e.g., do not fix
installation attachments on the flange of a beam but locate them on the web
near the neutral axis. In addition, keep bending moment arms short, locate
onsite joints of crane runway girders in a cross-section with the lowest bending
moment range, and so on.

Reduction of the structural stress concentration factor

Figure 16 shows various ways to reduce the factor K_s. Some improvement can
be achieved if the discontinuity is made smooth or gradual. However, the weld
is still located at the structural discontinuity, which means that both K_s and
M_k are, in effect, at the same point. The result will be better if the ends of the
attached parts are ground flush after welding. The best way to reduce the
effects of discontinuities is to move the weld away from the area where K_s is
effective. Thus, multiplication of two factors, K_s and M_k, at the same point

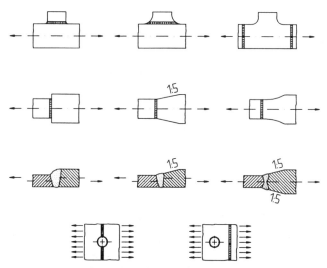

Fig 16 **Various ways of improving the design by reducing the structural stress concentration. The
best solution is to move the weld outside the stress concentration area**

can be avoided. Unfortunately, this very elegant design may be too expensive in many cases.

Offset and angular misalignments cause secondary plate bending stresses, Fig. 17. At critical locations such fabrication defects should be avoided. The angular misalignment can be reduced by minimizing the size of welds. Suitably strict tolerances should be specified if required.

Reduction of the non-linear stress peaks

The effect of non-linear stress peaks is reflected by the factor M_k in equation (1). In a transverse weld, high peak stresses are generated at the toes of large weld reinforcements. The small angle, θ, between the plate surface and the weld, and the large joint width, L, are the main reasons, see Fig. 18. Thus, excessive welding should be avoided, and a welding method which produces a smooth transition between the plate surface and the weld should be specified.

The effect of weld reinforcement is quite similar to the effect of an initial crack of a certain depth on the surface of a smooth plate, see Fig. 19. The larger and steeper the weld, the deeper the equivalent initial crack it represents.

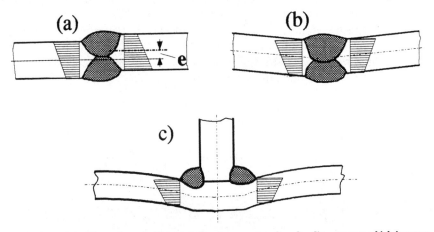

Fig 17   Offset (a) and angular misalignments (b, c) cause secondary bending stresses which increase the structural stress

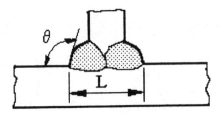

Fig 18   Excessive welding should be avoided. A small angle and a large width increase the peak stresses

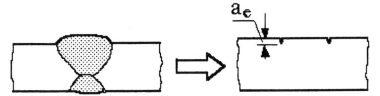

Fig 19   The effect of a transverse weld reinforcement could be substituted by two initial cracks of a certain equivalent depth

In a longitudinal weld, the stop/restart points form potential crack initiation sites. Whenever possible, a continuous automatic welding process should be specified. Non-linear stress peaks may also be produced by notches in the rest of the thermally-cut edge between longitudinal fillet welds. When such notches are considered critical, full-penetration welds should be specified.

Reduction of the initial crack

It is well known that undercut and small slag or oxide inclusions at the weld toe act as small initial cracks, a_0, which means that there is no significant crack initiation life when the stress fluctuation lies above the fatigue threshold of the joint. According to some investigations (13)–(15), the depth of the initial crack varies from 0.05 to 0.4 mm. A value of 0.25 mm is normally used in (1), leading to suitably conservative life estimations.

Welding procedures that produce a smooth transition between the base plate and the weld usually decrease the initial crack size as well. Poor transitions can be dressed after welding using either grinding or a suitable remelting process (16)(17). Remelting using either TIG or a plasma method is preferred because of the higher productivity. Usually, only small parts of weld ends need to be dressed.

Weld toe dressing has a twofold effect. Non-linear stress peaks will be reduced and initial crack removed.

Enlargement of the final crack

A crack propagates very rapidly in the last stage before final fracture, Fig. 12. Therefore, the crack propagation life becomes only a little longer when a_f is enlarged by choosing a material with a higher fracture toughness (18). In spite of this, a good toughness should be specified in order to make it possible to detect the growing crack well before it becomes critical.

Lowering of the stress ratio

In a welded structure, high tensile residual stresses increase the actual stress intensity factor ratio, R. A high value of R increases the paremeter $C = C(R)$ in the Paris' relation which means that the crack will propagate faster. Furthermore, and more importantly, a high R ratio lowers the fatigue threshold in the high cycle region, Fig. 20.

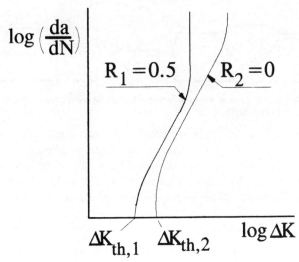

Fig 20 **A high true stress intensity factor ratio, *R*, increases the crack growth rate and lowers the threshold stress intensity factor**

The means of relieving the residual stresses, or sometimes even producing compressive stresses at the hot spot, are as follows:

(1) post weld heat treatment;
(2) hammering or shot peening;
(3) tensile overloading;
(4) vibratory stress relief.

The effect of stress relieving is best in cases in which the nominal mean stress level is low.

Quality assurance aspects

When the designer specifies special welding procedures, weld toe grinding or dressing, stress relieving or similar fatigue life improvement procedures, it is important that the instructions are understood correctly. The manufacturer's quality system manuals should include work instruction sheets which the designer can refer to. Otherwise, such instructions should be submitted by the designer. The results of the applied improvement procedure should be examined to a similar extent as the welding itself.

Fatigue loading considerations

The fatigue analysis of welded components should be based on characteristic stress range data. For some types of structures the applicable design codes specify the stress range design spectra, Fig. 21. However, in many cases such

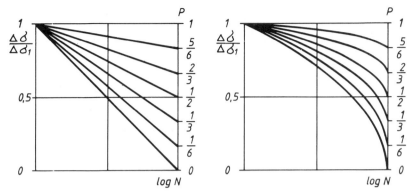

Fig 21 An example of a set of standardized stress range spectra (3)

data are not available. The designer must then solve the problem either by analysis or by field measurements from existing structures or prototypes. Analytical solution of the stresses resulting from multiple non-proportional moving loads is often impossible. Field measurements are then recommended.

From each field measurement period, the designer obtains a list of Rainflow-counted stress ranges and the numbers of their occurrences (graphical presentations can be seen in Fig. 23). The total range of the stress is divided into a certain number of bins which contain the numbers of occurrences found by the counting procedure. Each list represents a certain duration of use, which may be expressed as the number of loads lifted, length of a trip in kilometres, time in minutes, or any other suitable reference unit. Of course, the design should be based on a representative mix of measurements made under different operational conditions.

Usually, the list of stress range occurrences contains high numbers of small stress fluctuations. Therefore, the number of recorded stress cycles is not a good measure of the duration of operation. A better measure would be the number of loads, kilometres, minutes or other similar reference units. Instead of the direct application of Miner's rule, an equivalent constant amplitude loading may be used substituting for the variable amplitude loading

$$\Delta\sigma_{eq} = \left\{\frac{\sum_{i=j}^{k}(n_i\,\Delta\sigma_i^3)}{N_{ref}}\right\}^{1/3} \tag{2}$$

where

$\Delta\sigma_{eq}$ is a constant amplitude stress range;
N_{ref} is the number of reference units representing the duration of the variable amplitude loading sample;
n_i is the number of cycles in the sample in bin i;
$\Delta\sigma_i$ is the stress range corresponding bin i;
j is the lowest stress range bin causing crack growth;
k is the stress range bin containing the largest stress range.

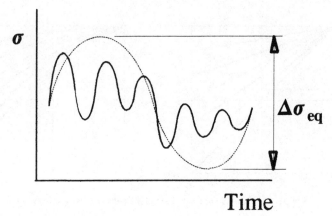

Fig 22 A single equivalent stress cycle which substitutes for the variable amplitude stress ranges during one working cycle

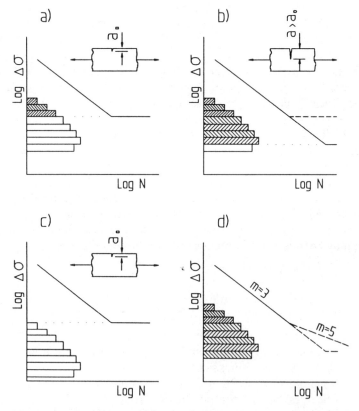

Fig 23 Effect of crack propagation on fatigue threshold (a, b). No propagation will occur in case (c). A reduced S–N curve slope is often used as a substitute for the changing threshold (d)

The derivation of equation (2) is presented in the Appendix, Fig. 22 explains the concept. The calculated $\Delta\sigma_{eq}$ is used for the determination of the expected fatigue life, either using a suitable S–N curve, or equation (1) based on fracture mechanics. In both cases the result gives the life expressed in reference units.

One problem in the analysis of components subjected to variable amplitude loading is the truncation of the small stress ranges. There is no fixed threshold if even only a small part of stress ranges exceed the fatigue limit found in constant amplitude testing. In such a case, some crack propagation will occur and the stress intensity factor ranges will gradually increase. Thus, an increasing number of stress ranges become effective, as shown in Fig. 23. When equation (1) is integrated step by step, it is quite easy to realize such a truncation if a relevant threshold of the stress intensity factor range is known. When the analysis is based on an S–N curve, this phenomenon is taken into account in several fatigue design codes by reducing the slope of the S–N curve after the endurance limit, also shown in Fig. 23.

Conclusions

The increasing use of the finite element method, and the growing understanding of the fatigue phenomenon through fracture mechanics, provide new opportunities for the designer of fatigue-loaded welded components. Notwithstanding the value of the accumulated knowledge gained by trial and error, good design practice should nowadays be based on a deep insight into the stress fields generated by structural discontinuities. The classified details of the design codes serve as a first screening of design alternatives, but they are not sufficient for details of relatively thin-walled high strength steel components, used, for example, in mobile equipment.

The so-called hot spot approach which is based on structural hot spot stresses has been developed for tubular joints in offshore structures. However, it has good prospects in many other fields, including ship hulls, mobile equipment etc. Parametric formulae for calculating the structural stress concentration factors for typical details in such fields are still to be developed.

Estimation of the crack propagation life of welded joints by fracture mechanics has already reached a maturity suitable for everyday use. The software employed must take into account the peak stress effects at weld toes and the development of the crack shape in a realistic way. The power of the fracture mechanics approach lies in the possibility of making various kinds of analyses required for a damage tolerant design and the possibility of comparing the merits of proposed design improvements.

An equivalent stress range is a good substitute for a direct application of Miner's rule. The resulting life prediction can then be expressed in chosen reference units instead of a number of stress cycles.

Appendix

Derivation of the fatigue-equivalent stress range

The S–N curve represented by a straight line in a log–log plot can be described by the following formula

$$N_f(\Delta\sigma)^m = C \tag{3}$$

where

N_f is the number of cycles to failure corresponding constant amplitude loading of magnitude $\Delta\sigma$;

m is a slope parameter;

C is the virtual value of fatigue life at $\Delta\sigma = 1$.

The partial damage, D_i, caused by the occurrences, n_i, at the stress range bin i, is written according to the well known Miner's rule

$$D_i = \frac{n_i}{N_{fi}} \tag{4}$$

Substituting N_{fi} from equation (3), and summing up all effective stress range bins yields the total damage

$$D = \frac{\sum_{i=j}^{k} (n_i \Delta\sigma_i^m)}{C} \tag{5}$$

where

j is the lowest bin causing crack growth;

k is the bin containing the largest stress range.

Defining an equivalent constant amplitude loading which would cause an equal amount of fatigue damage

$$D = \frac{N_{ref}(\Delta\sigma_{eq})^m}{C} \tag{6}$$

The number of occurrences, N_{ref}, can be chosen freely. Instead of the total number of cycles, Σn, it is convenient to choose the number of suitable reference units. A reference unit suitably describes the working cycle. Usual choices are the numbers of loading cycles, kilometres, or minutes. N_{ref} corresponds the same duration of work as the occurrences, n, in equation (5).

Substituting m by 3 which is a common value for as-welded structures, equalizing the damages from equations (5) and (6), and solving for $\Delta\sigma_{eq}$ yields the desired solution for the equivalent stress range

$$\Delta\sigma_{eq} = \left\{ \frac{\sum_{i=j}^{k} (n_i \Delta\sigma_i^3)}{N_{ref}} \right\}^{1/3} \tag{7}$$

Knowing the S–N curve, i.e., the constants C and m, and the equivalent stress range, the fatigue life (in the reference units) can be resolved from equation (3).

References

(1) Doc. IIS/IIW–693–81, (1982) Design recommendations, for cyclic loaded welded steel structures, *Welding in the world*, Vol. 20, pp. 163–165.
(2) Recommendations for the fatigue design of steel structures, (1985) ECCS Technical Committee 6 – Fatigue, First Edition.
(3) SFS 2378 (1985) Welding, Load capacity of welded joints in fatigue loaded steel structures, Suomen Standardisoimisliitto, p. 41 (in Finnish).
(4) GURNEY, T. R. (1979) Fatigue of welded structures, (Second edition), *Cambridge University Press*, Cambridge, 456.
(5) Doc. IIW–XIII–1458–92 (1992) Recommendations concerning stress determination for fatigue analysis of welded joints, draft proposal (Edited by E. Niemi), Lappeenranta University of Technology, p. 67.
(6) NIEMI, E. (1992) Determination of stresses for fatigue analysis of welded components, *Engineering Design of Welded Constructions*, Pergamon Press, Oxford, pp. 57–64.
(7) Fatigue Design Handbook (1968) *Society of Automotive Engineers*, Warrendale, p. 129.
(8) SILJANDER, O. A. (1991) Non-proportional biaxial fatigue of welded joints, PhD Thesis, Graduate College of the University of Illinois, Urbana IL.
(9) HUGILL, P. N. and SUMPTER, J. D. G. (1990) Fatigue life prediction at a ship deck/superstructure intersection, *Strain*, 107–112.
(10) WORDSWORTH, A. C. (1981) Stress concentrations at K, KT tubular joints, ICE Conference, Fatigue in offshore structural steel, paper No. 27.
(11) PETERSHAGEN, H., FRICKE, W., and MASSEL, T. (1991) Application of the local approach to the fatigue strength assessment of welded structures in ships, Doc. IIW–XIII–1409–91.
(12) HOBBACHER, A., (1988) Recommendations for assessment of weld imperfections in respect to fatigue, IIW–Doc. XIII–1266–88.
(13) YAMADA, K. and HIRT, M. A. (1982) Fatigue crack propagation from fillet weld toes, *J. Structural Division*, **108**, 1526–1540.
(14) PETERSHAGEN, H. (1984) The influence of undercut on the fatigue strength of welds – a literature survey, Doc. IIW–XIII–1120–84.
(15) BOKALRUD, T. and KARLSEN, A. (1982) Control of fatigue failure in ship hulls by ultrasonic inspection, Norwegian Maritime Research, No. 1.
(16) KNIGHT, J. W. (1977) Improving the fatigue strength of fillet welded joints by grinding and peening, Doc. IIW–XIII–851–77.
(17) KADO, S. *et al.* Fatigue strength improvement of welded joints by plasma arc dressing, Doc. IIW–XIII–774–75.
(18) HAUSAMMAN, H. (1980) *Influence of fracture toughness on fatigue life of steel bridges*, Lehigh University, Bethlehem Pa.

Index